DISCRETE MATHEMATICS
FOR COMPUTING

JOHN E MUNRO

CHAPMAN & HALL
London · New York · Tokyo · Melbourne · Madras

T.S.

Published by
Chapman & Hall, 2–6 Boundary Row, London SE1 8HN

Chapman & Hall, 2–6 Boundary Row, London SE1 8HN, UK

Blackie Academic & Professional, Wester Cleddens Road, Bishopbriggs, Glasgow G64 2NZ, UK

Chapman & Hall, 29 West 35th Street, New York NY10001, USA

Chapman & Hall Japan, Thomson Publishing Japan, Hirakawacho Nemoto Building, 6F, 1–7–11 Hirakawa-cho, Chiyoda-ku, Tokyo 102, Japan

Chapman & Hall Australia, Thomas Nelson Australia, 102 Dodds Street, South Melbourne, Victoria 3205, Australia

Chapman & Hall India, R. Seshadri, 32 Second Main Road, CIT East, Madras 600 035, India

First edition 1992

© 1992 John E. M. Munro   *f tw*

Printed in Australia   *ADS6308*

ISBN 0 412 45650 8

Cover design by Robert Bertagni

Contents

PREFACE   vii

1   Algorithms for Integers   1
   1.1   INTEGER DIVISION   2
   1.2   PRIMES   7
   1.3   GREATEST COMMON DIVISOR   10
   1.4   EXERCISES   13

2   Numbers in Different Bases; Real Numbers   17
   2.1   INTEGERS IN DIFFERENT BASES   18
       2.1.1   Changing base   19
       2.1.2   Arithmetic operations on positive integers   21
       2.1.3   Representation of integers in a computer . .   25
   2.2   REAL NUMBERS   30
       2.2.1   Rational numbers   30
       2.2.2   Irrational numbers   35
   2.3   EXERCISES   37

3   Sets, Functions and Relations   41
   3.1   SETS   41
       3.1.1   Properties of sets   44
       3.1.2   Set operations   44
   3.2   FUNCTIONS   48
   3.3   RELATIONS   54
       3.3.1   Equivalence relations   56
       3.3.2   Partial order relations   58
   3.4   SEQUENCES   59
       3.4.1   Strings   61
   3.5   EXERCISES   63

4 Logic   **69**

4.1 COMPOUND PROPOSITIONS & TRUTH TABLES   70

4.2 REASONING FROM LOGICAL ASSUMPTIONS   78

     4.2.1 Laws of logical algebra or propositional calculus 80

     4.2.2 Rules for replacement and transitivity . . . 81

4.3 PREDICATES & QUANTIFIERS 82

4.4 EXERCISES 86

5 Proof   **89**

5.1 PATTERNS OF PROOF 90

5.2 MATHEMATICAL INDUCTION 93

5.3 INDUCTION & WELL-ORDERING 99

     5.3.1 The principle of mathematical induction: Weak form   100

     5.3.2 The principle of mathematical induction: Strong form   101

     5.3.3 Well-ordering 103

5.4 EXERCISES 105

6 Recursion   **109**

6.1 RECURSIVE THINKING 110

     6.1.1 Recursive definition 110

     6.1.2 Problem solving 112

6.2 RECURSION & LISTS 118

6.3 RECURSION & SEQUENCES 120

     6.3.1 Calculation of the terms of a sequence . . . 122

     6.3.2 Solving a recurrence relation by observing a pattern   123

     6.3.3 Second order, linear, homogeneous recurrence relations 124

6.4 EXERCISES 127

7 Analysis of Algorithms   **133**

7.1 SLOW ALGORITHMS   133

     7.1.1 The Towers of Hanoi 133

     7.1.2 The travelling salesman 135

7.2 COMPARING ALGORITHMS   136

7.3 APPLICATIONS 142

     7.3.1 Searching 142

     7.3.2 Sorting 144

   7.4   EXERCISES 150

8 Graphs and Trees   **155**
   8.1   GRAPHS 155
       8.1.1   Matrices and matrix multiplication 160
       8.1.2   Description of a graph 162
       8.1.3   Walks, paths, and other connections 163
   8.2   TREES 166
       8.2.1   Connected graphs 167
       8.2.2   Trees 170
       8.2.3   Binary trees 172
   8.3   SPANNING TREES 175
       8.3.1   Search trees 176
       8.3.2   Minimum weight spanning tree 181
   8.4   EXERCISES 183

9 Counting   **187**
   9.1   TECHNIQUES OF COUNTING 187
       9.1.1   Numbers and counting 187
       9.1.2   Counting by tally 188
       9.1.3   Counting members of a sequence 189
       9.1.4   Counting Cartesian products 190
       9.1.5   Counting subsets 191
       9.1.6   The pigeonhole principle 192
   9.2   COMBINATORIAL REASONING 195
       9.2.1   A model for counting 195
       9.2.2   Combinatorial reasoning 199
   9.3   THE BINOMIAL & MULTINOMIAL THEOREMS 203
       9.3.1   The binomial theorem 203
       9.3.2   The multinomial theorem 205
   9.4   ALGORITHMS 207
       9.4.1   Subsets of a set 207
       9.4.2   Permutations 210
   9.5   EXERCISES 216

10 Algebraic Structures   **221**
   10.1 ALGEBRAS 221
       10.1.1 Operations and their algebraic properties . 222
       10.1.2 Algebras with one operation 230
       10.1.3 Groups 232
   10.2 ORDERED STRUCTURES 236

10.2.1 Partially ordered sets and lattices 236

10.2.2 Boolean algebras 240

10.2.3 Boolean algebras and lattices 242

10.3 FINITE FIELDS 244

10.4 EXERCISES 246

11 Reasoning and Correctness   **249**

11.1 NATURAL DEDUCTION 250

11.2 CORRECTNESS OF ALGORITHMS 260

11.3 EXERCISES 268

APPENDICES

A   Writing Algorithms   **271**

B   Answers to Exercises   **275**

Bibliography   **299**

Index   **301**

Preface

This book is intended for first year undergraduate students of computer science and information systems and for mathematics majors, although it has also been used by graduates enrolled in a computing diploma. A mathematical experience is now recognised as essential for the education of computing professionals and has an important place at the start of an undergraduate course. However students in computer science and information systems are a distinctly different group to the physicists, engineers, economists and others who are quite well served by the traditional first year undergraduate mathematics courses based on calculus and linear algebra. Topics such as software quality, artificial intelligence and others require a rigorous and early mathematical foundation set in the context of mathematical structures directly applicable to computing. Some of the early attempts to provide for computing students were too late in the course, or if given early were too elementary.

The book is the outcome of course development at the University of Canberra begun in 1985 and revisions based on the experience of giving the subject in nine successive semesters.

The subject 'discrete mathematics for computing' took on a life of its own during the 1980's. Conferences and committees laboured over the exchange of ideas and the construction of a suitable syllabus. The debate involved the three communities of computing logic and mathematics. There is a consensus emerging and this book is close to the general view on content.

For a sound mathematical foundation the main objectives are: to give fluency in precise mathematical language to deal with discrete structures, to develop skills in algorithmic problem solving, recursive thinking and combinatorial thinking, to develop respect for careful definition and skills in mathematical argument and to provide a knowledge of logic including truth tables, predicates, quantifiers and inference schemes.

Mathematics majors would also benefit from a serious look at the beginnings of discrete mathematics early in the course. The areas of discrete mathematics and computer mathematics are fruitful areas for the next generation to investigate. A course in discrete mathematics is an important complement to a course in calculus and linear algebra for those students looking for a broad view of modern mathematics. I hope this book may be used by students in both computing and mathematics.

Principles

The topics should be accessible but thought provoking. A serious attempt has been made to limit the size of the book by taking (sometimes difficult) decisions on what mathematics is most likely to be valuable in computing. Undoubtedly some lecturers will be disappointed that particular topics are not here, but from the student point of view the approach makes for a less voluminous, more focussed book.

The needed skills can be brought to a useful level in one semester using a fairly sparing approach to the inclusion of topics. Material should be learnt at a fairly high mastery level, with the expectation that skills will be used, algorithms constructed, proofs devised and predicates stated clearly. This should be a course that develops new modes of thinking not merely a study of the application of recipes in a routine fashion. At the University of Canberra our experience has been that students vary widely in both mathematical background and age. Some come to change or broaden an existing career. Others come directly from school. The required background is secondary school algebra, to the level of what is often called 'pre calculus'. Calculus itself is not necessary. However it is important to be willing to think deeply about the subject. There are ideas involved that will test even the best prepared.

There are two approaches to motivation that are appealing but take us the wrong way.

It is a mistake to think that simplification and 'apparent success' will satisfactorily motivate learning. The proper role for simplification in mathematics is to give insight so that real, usually harder or more complex problems can be solved. What is needed is to develop strategies to deal with new problems, some of which will be difficult. The achievement of success in this context provides lasting motivation.

The second approach to motivation that I have difficulty with is the excessive pursuit of applications. The inclusion of algorithms in computer code, the presentation of computer engineering solutions to problems, for example the design of a 'half-adder' and so on, seem to me to pursue this aspect of motivation beyond what is reasonable. Although students like some indication of the places where mathematics is used, algorithms in computer code and specifically computing topics are better placed in a computing course.

As much as possible of the time available should be spent on the development of the central mathematical ideas in the areas of mathematical argument, recursive methods and the language of sets relations and functions. The important ideas have a multitude of different settings and variations. Students must meet these ideas in different situations and experience a *variety* of exercises before the knowledge can be transferred to solve new problems in new situations. For students to be able to call on the central concepts and skills with confidence, much more time is needed than appears on the lecture schedule so that the transfer will work. The purpose behind including applications, where this is done, is to help students in the process of transfer.

A potential problem for students is an apparent lack of connection between some topics. Because the book really deals with the beginnings of quite a number of well established branches of knowledge it is heavy with definitions and concepts. To combat this problem the material has been integrated with common structures such as relations and functions and a consistent approach to organising knowledge, using definitions, axioms, theorems, examples and proofs. Probably more useful from a student point of view is that the exercises present ideas over a longer period than the chapter headings would suggest. For example most sets involve some questions on mathematical reasoning although mathematical proof is dealt with formally in just one chapter. The emphasis throughout is on developing mathematical methods, language and strategies to solve worthwhile problems.

Content

Although the mathematical facts in Chapters 1 and 2 have value in themselves, much of the mathematical content should either be known already or easily learned. The choice of material is deliberate. A student should be free, in the early part of the course, to focus on learning how to produce *constructive* solutions to problems using language with sufficient precision for instructions to be followed automatically. The approach should be more formal than many students will expect.

Chapters 3, 4, 5 and 6 are about central mathematical skills and techniques of value in both computing and mathematics, namely sets, functions, relations, logic, proof and recursion.

Chapter 7 is a fairly informal study of the efficiency and correctness of algorithms designed partly to illustrate techniques developed earlier.

The next two chapters introduce graph theory and combinatorial mathematics. These are relatively modern areas of mathematics and have many applications in computing.

Chapter 10 is an introduction to algebraic systems. The first objective is to help in simplification by providing a language to describe structure. The second is to prove general results, which is economical because they have applications in more than one place.

The final chapter aims to introduce mathematical applications that look to the future. There are different opinions in the computing fraternity about the value of these topics in part because they are closer to areas of current research and so 'not proven'. They are in the areas of artificial intelligence and software engineering. The two topics introduced are *natural deduction* and *proof of correctness of algorithms.*

Acknowledgements

I wish to thank all those who have encouraged me in the development of introductory Discrete Mathematics for Computing at the University of Canberra; in particular Mary O'Kane, Peter O'Halloran, Brian Stone and Ivor Vivian. Many students, tutors and lecturers in computing and mathematics have discussed ideas and sections of the book. My sincere thanks to them all. I am very grateful to Roger Curnow, Neil Porter, Malcolm Brooks, Jan Newmarch, John Campbell and Brian Stone for detailed criticism of drafts. Deirdre Sheppard, John Campbell, Roger Curnow, Malcolm Brooks and Phillip Young have each taught versions of the material and, along with many students and tutors, have made invaluable suggestions for improving communication with the reader. I would like to thank Mirion Bearman, Van Le, Peter O'Halloran, Jan Newmarch, Donald Overheu, Arthur Zawadski, Robert Latta, Brian Goodwin, Peter Kenne, Clem Baker-Finch, Karen George, Errol Martin and Mike Newman for helpful information and for discussing ideas and correcting mistakes. It is impossible to identify all sources of ideas for a book such as this, in particular those anonymous reviewers arranged by the publisher. The reviewers did much more than say yes or no to particular questions about the proposal, in several cases giving very thoughtful responses to the draft. Ted Gannan of Thomas Nelson has been unfailing in his encouragement, which is very much appreciated. Janet Mau and Daphne Rawling have improved the book immensely by their careful editing. I wish to acknowledge the helpful ideas of E.R. Muller of Brock University Canada, who was among the pioneers of this kind of course in the early 1980s.

I have enjoyed writing the book for many reasons but perhaps most for the interaction with those who were kind enough to engage me in discussion and debate on the many issues and myriad details of the enterprise. They have all improved the result (even if we disagreed on particular issues). I hope they will be pleased with the product.

I would also like to thank The University of Canberra for giving me some time and The Australian National University for giving me a place to do the actual writing of the book.

John Munro

December 1991

Chapter 1

Algorithms for Integers

In this chapter we give algorithmic solutions to five problems involving integers. You should learn how to read and use these algorithms; know the facts presented on integers, primes and greatest common divisors, and remember what was said about preconditions and postconditions when you reach Chapter 4. At a higher level you should be able to read and use similar algorithms when you meet them and to design your own algorithms to solve problems about integers.

A modern electronic (digital) computer functions at the lowest level by switching currents on and off, changing voltages between two values or by reversing the magnetic polarity of pieces of magnetic material. We do not have to know much about how this is done, but at this basic level a component of the machine has two states which may be taken to represent the digits 0, 1, or the logical labels true(T), false(F), at our convenience. A mathematical structure that builds on this digital foundation is that of the integers, so we begin there. The natural way to instruct a computer to carry out a task is by means of an *algorithm*, by which we mean a step-by-step procedure for doing the task. For a computing student an algorithm stands midway between a problem to be solved and the program needed to solve it. For a mathematics major an algorithm represents a constructive solution to a problem as opposed to an argument that a solution exists. Davis and Hersh (1981) discuss the contrast between the algorithmic and the dialectic (argument based) approach for a mathematician in a very readable article.

However, combining algorithms and integers in the first chapter has a slightly deeper significance than the value of these subjects in themselves. Through most of secondary school an objective has been to embed integers (whole numbers) within the number system, using the same symbols for the operations of addition, multiplication, subtraction and division to emphasise the idea. A close look at the algorithms for performing these operations makes clear that there are different number systems which need to be handled in different ways by a machine such as a computer.

The algorithmic approach is found in the earliest mathematics, and has always been an important part of the subject. Algorithmic solutions are given to many of the problems presented in this book. Much of your early experience of mathematics is certain to have been algorithmic in character. But then your task would have been to conform to a teacher's instructions on, say, how to carry out a 'long' multiplication. Now you need to develop skills in producing your own algorithms to solve problems. This is a creative task. We hope that you will develop a style that is lucid, correct and efficient.

Algorithms are important to us because they give the kind of solution to a problem that can be converted into a program for a computer. In order to write a program or understand a program written by someone else, it is necessary to see its purpose and structure. An algorithm presents the structure of a program uncluttered by the necessary detail. The process of creating a program from an algorithm is called *implementation*.

There is, perhaps unfortunately, no universally accepted method of writing algorithms. In this book a form of language, sometimes called a 'pseudocode', is used. It is intended to be as close as possible to ordinary language. This special form of language is due to the fact that some basic instructions occur frequently in algorithms and are probably more easily recognised if presented in a standard form. In Appendix A some of the issues associated with writing algorithms are discussed.

Historical comment: The name algorithm derives from the Arab mathematician Mohammed al-Khwarismi (*c.* AD 825) who wrote a book on arithmetic which gave a full account of Hindu numerals. The book was translated into Latin (*De Numero Indorum*) and became known in Europe. The scheme of numeration using Hindu numerals came to be called *algorismi*, a corruption of al-Khwarismi. The meaning of algorithm is now 'any rule of procedure'. In computing most authors (e.g. Knuth, 1969, p.4)) emphasise that the application of the algorithm to valid input data should terminate in a finite number of steps.

1.1   INTEGER DIVISION

The set of integers, $\{..., -3, -2, -1, 0, 1, 2...\}$, denoted by $\mathbf{Z}$, is familiar. We know what is meant by the symbols "$=$", "$<$", "$\leq$", "$>$", "$\geq$" *defined on* the integers. For example we know quite definitely whether "$a = b$" is true or false for each pair of integers a,b in $\mathbf{Z}$. However, it may not be clear what is meant by division. Suppose a person or machine is instructed to divide a

by b.

For $a = 12, b = 3$ there is no doubt what is meant.

For $a = 13, b = 5$ it is not clear whether the answer 2.6 or 2 remainder 3 (or even 2 or 3) is wanted. 'The fundamental issue is not "what is correct" but "how should the intention be made clear".' One method is to define different symbols for each operation that is needed. Another method is to use the same symbol to mean different things by defining it to behave differently on different types of number. We will explore the task of providing precise instructions by looking at the division of integers.

The ordinary process of dividing an integer a by an integer b gives a quotient q and a remainder r. The facts are contained in the following theorem.

Theorem 1.1 (Division theorem) *For all integers a, b, $(b \neq 0)$, there exist unique integers q, r such that $a = b * q + r$ and $0 \leq r < |b|$.*

The convention in mathematical writing is not to use a special symbol, such as '$*$', to represent multiplication. So we will write 'b is multiplied by q' as 'bq'. Where necessary for clarity, '$*$' will be used for multiplication, for example '3 multiplied by 4' will be written $3 * 4$. The symbols '$|b|$' are used to indicate the absolute value of b, for example $|-7| = 7, \quad |7| = 7$.

It has been traditional to call the division theorem the 'division algorithm'. However the theorem does not identify a step by step procedure to obtain q, r from a, b, so in the context of this book it is not appropriate to call the theorem an algorithm.

Problem 1.1 *How should we calculate the remainder and the quotient for the division of one positive integer by another?*

There are various ways this may be done, depending on what functions are available on your calculator or computer. However the operation of subtraction is sure to be present on any machine that handles integers, so we base the following algorithm on the idea that division of integers corresponds to repeated subtraction.

Algorithm 1.1 (Positive integer division) *To calculate the remainder and quotient for the division of positive integer a by positive integer b.*

In the algorithm below the variables a, b are given initial values, by input from some source external to the algorithm, perhaps a keyboard entry or perhaps some computer program that has called up this algorithm, and a, b hold the initial values throughout the application of the algorithm. The variables q, r are assigned initial values during the application of the algorithm and the values may change repeatedly as the application continues. Each change is said to 'update' the value of the variable concerned.

input:	a, b, integer. $a > 0$ and $b > 0$.
output:	q, r, integer. The quotient q and the remainder r must satisfy the definition contained in theorem 1.1 above.
method:	1. Give q the value 0.
	2. Give r the value a.
	3. While $r \geq b$:
	4. add one to the value of q
	5. subtract b from the value of r.

The lines are numbered to help in the explanation of the working of the algorithm. It is not usual to number the lines in an algorithm. The instructions on line 1, line 2 are carried out one after the other in the obvious sequence. Note that by assigning these values to q, r we have satisfied part of the output condition $(a = bq + r)$. The next instruction is on line 3 and is a test to decide if r is outside the required range of values for the remainder. If the statement '$r \geq b$' is true, the indented instructions on line 4 and line 5 are performed and then the algorithm returns to the 'while' test, line 3. If the statement '$r \geq b$' is false, the algorithm terminates; otherwise the sequence line 4, line 5, line 3 is repeated. The indented instructions are referred to as a 'loop'. Each time the algorithm passes through the loop the value of r is reduced so that after a (finite) number of steps the test on line 3 must fail, and the algorithm terminates, with the correct values for quotient and remainder stored in q and r respectively.

Let us step through an example to observe the working of the algorithm.

Example: Find the quotient and remainder when 33 is divided by 7.

Identify the variables in algorithm 1.1 and list their values in a table as we step through the algorithm.

	a	b	q	r	$(r \geq b)$
initial values	33	7	0	33	T
after loop one			1	26	T
two			2	19	T
three			3	12	T
four			4	5	F

When 33 is divided by 7 the quotient is 4 and the remainder is 5.

The expression $(r \geq b)$ can take on the values true(T) or false(F) and no others. It is called a *Boolean* expression.

Problem 1.2 *What are the quotient and remainder when one integer is divided by another, there being no restriction on the sign of the integers?*

Four possibilities must be considered, taking a positive or negative with b positive or negative. Comments in italics have been placed in the algorithm to help understanding.

Algorithm 1.2 (Integer division) *To find the quotient and remainder when an integer a is divided by an integer b, where $b \neq 0$.*

input: a, b, integer, where $b \neq 0$.

output: q, r, integer. The quotient q and remainder r, must satisfy the condition $a = bq + r$ and $0 \leq r < |b|$.

method: Give q initial value 0.

        Give r initial value a.

        Case 1: $a \geq 0$ (*a positive or 0*)

           While $r \geq |b|$

                $\left\{\begin{array}{l} \text{replace } r \text{ by } r - |b| \\ \text{if } b > 0 \text{ then increase the value of } q \text{ by 1} \\ \qquad \text{else (if } b < 0 \text{ then) decrease } q \text{ by 1} \end{array}\right.$

        Case 2: $a < 0$

           While $r < 0$

                $\left\{\begin{array}{l} \text{replace } r \text{ by } r + |b| \\ \text{if } b > 0 \text{ then decrease } q \text{ by 1} \\ \qquad \text{else (if } b < 0 \text{ then) increase } q \text{ by 1} \end{array}\right.$

Example: Apply the integer division algorithm to -15 divided by -7.

The value of a is -15; the value of b is -7. The algorithm selects case 2 ($a < 0$) and, after the 'if', selects the 'else' branch ($b < 0$).

a	b	q	r	$(r < 0)$
-15	-7	0	-15	T
		1	-8	T
		2	-1	T
		3	6	F

When -15 is divided by -7 the quotient is 3 and the remainder is 6.

Algorithm 1.2 is more complicated than algorithm 1.1. It would make sense to check that it does indeed give the correct output for all legitimate inputs.

We introduce a statement called a *precondition* to check that the input is legitimate and a *postcondition* to check that the output is correct for a given input.

Given that a, b, q, r are integers, the precondition for the division algorithm is the expression '$b \neq 0$' and the postcondition for the division algorithm is the expression '$a = bq + r$ and $0 \leq r < |b|$', where q, r have their final output values. From now on for the division algorithm we write:

$$precondition(a, b) \quad = \quad (b \neq 0)$$
$$postcondition(a, b, q, r) \quad = \quad (a = bq + r \text{ and } 0 \leq r < |b|)$$

The *precondition* is tested on the input variables a, b before the algorithm is applied. The *postcondition* is tested using the values of the input variables that hold before the application of the algorithm and the values of the output variables q, r that hold after execution of the algorithm. Some care must be taken with the above explanation, because variables (e.g. q, r) change value during the execution of the algorithm. Each condition function is definitely true or definitely false at the time of its evaluation.

Now, if for each possible input pair a, b the *precondition* is true and after application of the algorithm the corresponding *postcondition* is true, then the algorithm must be correct. Of course we cannot check all of the infinitely many possible inputs. However, if at least we check the four possibilities a positive and negative with b positive and negative, our confidence in the algorithm will be increased. If the algorithm is *implemented* as a computer program, then the conditions can be evaluated automatically each time the program is run.

For the example above in which $a = -15, b = -7$:

$$precondition(-15, -7) \quad = \quad (-7 \neq 0)$$
$$= \quad \mathsf{T}$$
$$postcondition(-15, -7, 3, 6) \quad = \quad ((-15 = -7 * 3 + 6) \text{ and } (0 \leq 6 < 7))$$
$$= \quad \mathsf{T}$$

Example: The integer division algorithm produces output $q = -3, r = 6$ from the input $a = -15, b = 7$. Show that the algorithm has met its specification in this case.

We have to show that, if the input data satisfies the *precondition* and the algorithm produces the stated output, then the *postcondition* will be true. Thus:

$$precondition(-15, 7) \quad = \quad (7 \neq 0)$$
$$= \quad \mathsf{T}$$
$$postcondition(-15, 7, -3, 6) \quad = \quad ((-15 = 7 * (-3) + 6) \text{ and } (0 \leq 6 < 7))$$
$$= \quad \mathsf{T}$$

Note 1: Ultimately we would like to be able to prove that the algorithm for integer division is correct for all valid inputs. That is, for all integers a, b:

if *precondition(a,b)* is true and the algorithm applied to (a, b) produces the output (q, r) then *postcondition(a,b,q,r)* is true.

Towards this goal we will study logic and mathematical proof in Chapters 4 and 5.

Note 2: Some computer languages contain operators that calculate q and r automatically. For example, in Pascal q is evaluated by 'a div b' and r is evaluated by 'a mod b.'

1.2 PRIMES

Definition 1.1 (Divisible) *An integer b is said to be* divisible *by an integer a if and only if $b = qa$ for some integer q. We say a is a* factor *of b.*

Definition 1.2 (Prime) *A positive integer p is* prime *if p is not 1 and if p is divisible only by 1 and p.*

Positive integers greater than 1 that are not prime are called *composite*. The first few positive primes are 2, 3, 5, 7, 11, 13, 17, 19, 23,... Any positive integer greater than 1 is either prime or can be factored into primes, e.g.

$$128 = 2^7, \ 45 = 3^2 * 5.$$

From experience we know that the prime factors of a number are always the same no matter how we set about the process of factorisation. It is possible to prove that the factorisation of a composite into primes is unique.

Theorem 1.2 (Fundamental theorem of arithmetic) *Every positive integer greater than 1 can be expressed as a product of positive primes. This expression is unique apart from order.*

The phrase 'product of primes' is taken to include the possibility of a single prime.

The problem below is a good example of a *decision problem*. Problems of this type occur frequently. Since the aim is to determine whether a given statement is true or false, a Boolean variable (in this case the variable *prime*) is often defined in an algorithm for a decision problem.

Problem 1.3 *Is some given positive integer, n, prime?*

This seems too simple a question to require the use of a computer; but ask 'is 21 509 a prime?' or 'is 432 521 a prime?' Even using a calculator it helps to organise the sequence of calculations to minimise the work. We could test n for divisibility until a factor is found or all possible factors have been tried. However each test for divisibility gives two factors.

If 2 is not a factor then $n/2$ is not a factor. Any possible factors now lie between 2 and $n/2$. The interval containing possible factors shrinks as each test fails, until the last possibility is to test for $\sqrt{n}$, if that number is an integer.

The sequence of possible factors to try is therefore $2, 3, \ldots, \lfloor \sqrt{n} \rfloor$, and if all these fail we are sure that the number is prime.

The symbols $\lfloor \quad \rfloor$ denote the *floor* function which is defined by

$$floor(x) = \lfloor x \rfloor = \begin{cases} x, & \text{if } x \text{ is integer} \\ \text{greatest integer less than } x, & \text{otherwise} \end{cases}$$

Examples: $\lfloor 3.9 \rfloor = 3,$ $\lfloor -3.2 \rfloor = -4,$ $\lfloor 5 \rfloor = 5.$

Algorithm 1.3 (Prime test) *To test whether an integer n, is prime.*

> **input:** n, integer. The integer to be tested.
>   A precondition is $(n > 2)$.
> **output:** *prime*, Boolean.
>
> The postcondition is $(prime = \begin{cases} \mathsf{T} & \text{if } n \text{ is prime,} \\ \mathsf{F} & \text{otherwise.} \end{cases})$
>
> **method:** Give a the value 2.
>   *(a is the integer being tested as a possible factor)*
>   Give *prime* the value T.
>   While $a \leq \sqrt{n}$ and *prime* is T:
> $\begin{cases} \text{if } a \text{ divides n exactly} \\ \qquad \text{then give } prime \text{ the value } \mathsf{F} \\ \qquad \text{else increase the value of } a \text{ by 1} \end{cases}$

Is the algorithm correct? efficient? easily understood?

Note that the above method would fail for $n = 2$ and that value has been excluded by the precondition. Another approach would be to change the method to deal with $n = 2$ as a possible input. An algorithm may be shown to be correct by establishing that it gives the correct results for all inputs. It may be proved incorrect by showing that it fails to meet the specifications for even one valid input. The question of correctness is examined more deeply in Chapter 11.

The question of efficiency is looked at in Chapter 7.

The question of readability is to some extent subjective, although of great importance in professional computing. You will see algorithms written in a variety of ways in different books, and so you need to be able to read different styles. Some notes on writing algorithms are given in Appendix A.

Problem 1.4 *What are the prime factors of an integer greater than 1?*

For any positive integer n greater than 1 we wish to produce a list of the prime factors of n, repeating factors where appropriate; for example, 12 has factors $2, 2, 3$. It is necessary to test n for divisibility by the numbers from 2 to $\sqrt{n}$ in order. However, when a factor is found, continue to divide by that factor while it succeeds as a factor. This procedure ensures that each factor found is prime. For example, 12 is not found as a *prime* factor of 42, since all factors 2 and 3 will have been removed before the test for 6 is applied.

The variables required for this algorithm are 'the trial divisor', denoted by a, and 'the quotient left after dividing out a successful trial divisor', denoted by q. In computing the practice is to represent such variables by a self-explanatory name such as *trialdivisor* or *quotient*.

Algorithm 1.4 (Prime factors) *To find the prime factors of an integer greater than 1.*

The algorithm for prime factors is described below.

A Nassi-Schneidermann diagram (favoured by some writers) for the same algorithm is shown in Figure 1.1 below.

> **input:** n, integer. The precondition is $(n > 1)$.
>
> **output:** A list of integers $f = < f_1, \cdots f_k >$.
> The postcondition is: (each of $f_1, f_2, ... f_k$ is prime, and $f_1 * f_2 * ... * f_k = n$.)
> *Note: f_j, f_k are not necessarily distinct.*
>
> **method:** Let f be $<>$ *($<>$ is the empty list)*
> Assign q the value n.
>   *(q is the quotient after the most recent division)*
> Assign a the value 2.
>   *(a is the current trial divisor)*
> While $a \le \sqrt{q}$:
>   if q is divisible by a
>   then $\left\{ \begin{array}{l} \text{append } a \text{ to } f \\ \text{assign } q \text{ the value } q/a \end{array} \right\}$ branch 1
>   else $\{$ increase a by 1 $\}$ branch 2
> append q to f

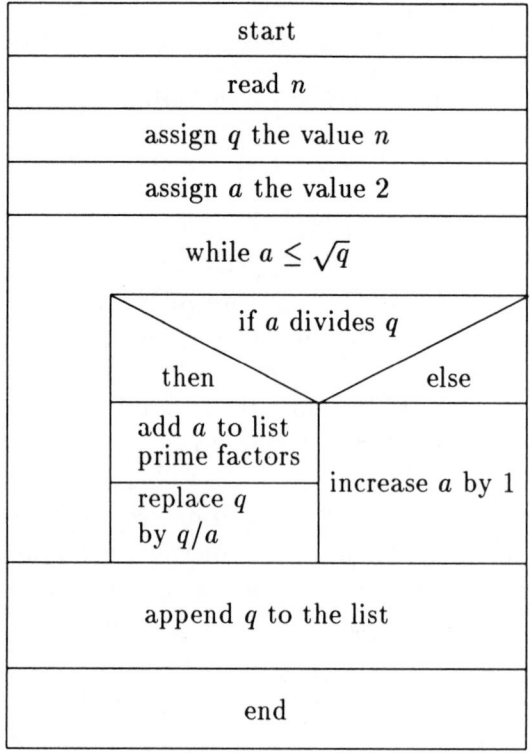

Figure 1.1: *Nassi-Schneidermann* diagram

Example: Find the prime factors of 60 using algorithm 1.4.

n	f	q	a	$(a \leq \sqrt{q})$	branch
60	$<>$	60	2	T	branch 1
	$< 2 >$	30	2	T	branch 1
	$< 2,2 >$	15	2	T	branch 2
	$< 2,2 >$	15	3	T	branch 1
	$< 2,2,3 >$	5	3	F	exit loop

$$f = < 2,2,3,5 >$$

1.3 GREATEST COMMON DIVISOR

Historical comment: Euclid lived about 300 BC. He organised classic Greek mathematics into thirteen books called *The Elements*. His algorithm for greatest common divisor, given below, is in Book VII. Euclid established the tradition of building a logical sequence of theorems or propositions on a foundation of axioms and definitions. *The Elements* and the axiomatic reasoning system used in it had an immense influence on the development of mathematical thought, particularly *mathematical proof.*

The greatest common divisor of two integers m, n is the largest integer that is a factor of both m and n. The greatest common divisor is defined for all pairs of integers that are not both zero. The usual notation is $\gcd(m, n)$. For example

$$(1) \ \gcd(91, 13) = 13, \quad (2) \ \gcd(-3, -6) = 3, \quad (3) \ \gcd(8, 0) = 0.$$

Problem 1.5 *What is the greatest common divisor of two integers m and n, not both zero?*

If $m \neq 0$, $\gcd(m, 0) = |m| = \gcd(0, m)$.

If $m \neq 0$ and $n \neq 0$ then $\gcd(m, n) = \gcd(n, m) = \gcd(|m|, |n|)$. So it will be enough to find an algorithm for $\gcd(m, n)$ where $m \geq n > 0$.

A feasible approach to this problem would be to compare prime factorisations of m and n, obtained by using algorithm 1.4 above, but there is a much more efficient algorithm available.

Algorithm 1.5 (Euclid) *To find the greatest common divisor of two positive integers.*

    input: m, n, integer.
                The precondition is ($m \geq n > 0$).
    output: d, integer.
                The postcondition is (d divides m and d divides n and if c divides both m and n then c divides d).
    method: Give a the value m, give b the value n.
                Calculate r, the remainder when a is divided by b.
                While $r \neq 0$:
                      assign a the value of b
                      assign b the value of r
                      recalculate r, the remainder when a is divided by b
                Give d the value of b.

Example: Find the greatest common divisor of 2436 and 1015.

    Step through the algorithm:

a	b	r	$(r = 0)$
2436	1015	406	F
1015	406	203	F
406	203	0	T

$$\text{thus}\ \ \gcd(2436, 1015) = 203$$

Some computer languages have a function to evaluate the remainder when one integer is divided by another. For example the function *mod* is defined for integers a, b where $a \geq 0$ and $b > 0$ by

$$a \bmod b = \text{ the integer remainder when } a \text{ is divided by } b$$

(We say 'a modulo b' or '$a \bmod b$'.)

Using the *mod* function the above calculation becomes

$$
\begin{aligned}
2436 \bmod 1015 &= 406 \\
1015 \bmod 406 &= 203 \\
406 \bmod 203 &= 0
\end{aligned}
$$

The postcondition for Euclid's algorithm

The postcondition for the Euclidean algorithm is difficult to apply directly. Although the output d is easily checked to be *one* common divisor, how do we know that it is the greatest one? How do we prove that the algorithm terminates correctly? The key is that at the end of each passage through the loop, the value of $\gcd(a, b)$ is unchanged, although a, b themselves have changed. The identification of such *invariant* properties are important in proving that loops operate correctly. The traditional proof for Euclid's algorithm follows. The invariance property of $\gcd(a, b)$ is established in the second part of the proof.

Proof of Euclid's algorithm for greatest common divisor

The proof of an algorithm should convince you that the algorithm terminates correctly in a finite number of steps.

The algorithm requires a sequence of integer divisions(step 1,3). Denote successive remainders by $r_1, r_2, \ldots$ and quotients by $q_1, q_2, \ldots$ By this means we avoid the possible confusion that might be caused by the fact that a, b change value through the operation of the algorithm.

To prove the algorithm correct two things are required. It must be shown that the algorithm terminates; and it must be shown that at termination the correct 'greatest common divisor' is obtained.

1. Termination Show that the algorithm terminates.

By the division theorem(theorem 1.1) we have:

$$
\begin{aligned}
m &= n * q_1 + r_1 \text{ and } 0 \leq r_1 < n \qquad &(1.1)\\
n &= r_1 * q_2 + r_2 \text{ and } 0 \leq r_2 < r_1 \\
r_1 &= r_2 * q_3 + r_3 \text{ and } 0 \leq r_3 < r_2 \\
&\quad \cdots \\
r_{j-2} &= r_{j-1} * q_j + 0
\end{aligned}
$$

Taking the inequalities together we have $n > r_1 > r_2 \cdots$, so this *decreasing* sequence of *non-negative* integers must reach zero. Suppose this happens after j divisions. Then $r_j = 0$, and the algorithm *terminates* with the variable d equal to r_{j-1}.

2. **Correct termination** Show that r_{j-1} is the greatest common divisor of m and n.

   (The proof depends on showing that, although a, b change value during the execution of the algorithm, $\gcd(a, b)$ returns to the same value at the end of each pass through the loop. We say that this variable is a *loop invariant*.)

   Since $\gcd(m, n)$ divides both m and n, by equation (1.1) $\gcd(m, n)$ also divides r_1. Thus $\gcd(m, n)$ divides $\gcd(n, r_1)$.
   Similarly equation (1.1) shows that $\gcd(n, r_1)$ divides m and hence divides $\gcd(m, n)$.

$$
\begin{aligned}
\text{Therefore } \gcd(m, n) &= \gcd(n, r_1) \\
&= \gcd(r_1, r_2) \\
&= \cdots \\
&= \gcd(r_{j-1}, r_j) \text{ where } r_j = 0 \\
&= r_{j-1}
\end{aligned}
$$

   which shows that the algorithm terminates correctly.

1.4   EXERCISES

1. Use algorithm 1.2 to find the quotient and remainder when:

   (a) 18 is divided by 7
   (b) 18 is divided by -7
   (c) -18 is divided by 7

   In each case evaluate the *precondition* and the *postcondition*.

2. What is the largest number that should be tested as a possible factor before it is certain that 1987 is a prime?

3. Determine whether 11729 is prime.

4. A perfect number is one for which the sum of all the factors of the number smaller than itself equals the number. Thus 6 is a perfect

number. For 1, 2, 3 are the factors of 6 other than 6 itself; and
$1 + 2 + 3 = 6$.

Construct an algorithm to test whether an integer is perfect.

Step through your algorithm for $n = 24$ and $n = 28$.

5. Find the greatest common divisor of 3589 and 999 using the Euclidean algorithm. Given that an upper limit to the number of divisions required is

$$\lfloor 2 \log_2 999 \rfloor$$

calculate this upper limit.

6. Use the prime factor algorithm 1.4 to find the prime factors of 61 and of 64. Which of these inputs makes most use of the *while* loop?

7. Let m be a positive integer greater than one. Then the integers $0, 1, \ldots, (m-1)$ together with the operations for addition and multiplication defined below form a number system called $\mathbf{Z}_m$.

The *addition* of a pair of numbers x, y in $\mathbf{Z}_m$ is defined to be r, the remainder on dividing $(x + y)$ by m. We write

$$x + y \equiv r \bmod m$$

and say $x + y$ is congruent to r modulo m. For example $\mathbf{Z}_{11}$ consists of the integers $0, 1, 2, \ldots, 10$; and

$$3 + 5 \equiv 8 \bmod 11,$$

$$7 + 9 \equiv 5 \bmod 11.$$

The *multiplication* of a pair of numbers x, y in $\mathbf{Z}_m$ is defined to be s, the remainder on dividing xy by m. For example

$$3 \times 2 \equiv 6 \bmod 11,$$

$$7 \times 9 \equiv 8 \bmod 11,$$

where 8 is the remainder on dividing 63 by 11.

Construct an algorithm using subtraction of integers, for the addition of numbers in $\mathbf{Z}_7$. The algorithm should accept any pair of 'number's in $\mathbf{Z}_7$ and output a 'number' from $\mathbf{Z}_7$.

8. Euclid's algorithm 1.5 uses integer division.

Division may be carried out by repeated subtraction (algorithm 1.1).

Design a greatest common divisor algorithm that uses subtraction directly.

9. Read Donald Knuth (1969, pp. 1-9) on algorithms.

10. Read Philip Davis and Reuben Hersh (1981, pp. 180-6).

Chapter 2

Numbers in Different Bases; Real Numbers

The representation of integers in bases other than 10, particularly 2,8 and 16, is the first topic in this chapter. Examples are given to show how the everyday algorithms for addition, multiplication and so on are modified when positive integers are expressed in non-decimal bases. Next we look at one way in which integers are represented in a computer. Finally real numbers(which include fractions) are discussed. Although the same symbols are used for operations on real numbers and on integers, the underlying algorithms are necessarily different. Algorithms are presented to show how the base system and arithmetic operations work. You should learn to use the algorithms and know the facts about bases and real numbers. At a higher level you should be able to design algorithms to solve problems and use previously unseen algorithms designed by others.

In ordinary life, numbers are represented using ten digits. The position of each digit determines its value on a base of ten. For example, the string of digits 325 is a shorthand for

$$300 + 20 + 5$$

or

$$3 \times 10^2 + 2 \times 10^1 + 5 \times 10^0$$

However it is not necessary for the base to be ten, and for hardware reasons base 2(binary), base 8(octal) and base 16(hexadecimal) are frequently used in computers. Mathematical reasons that support those choices of base are first, that the multiplication and addition tables in binary have just four entries, by contrast with the hundred required for base 10; and second, the algorithms for changing between these bases are quite efficient.

2.1   INTEGERS IN DIFFERENT BASES

In this book, 'a non-zero integer is represented in base x' means 'the integer is written in the form

$$sa_k a_{k-1} \ldots a_1 a_0 \quad (\text{base } x)$$

where s, the sign of the integer is $+$ or $-$; where a_i is an integer such that $a_k \neq 0$ and for all other i, $\quad 0 \leq a_i < x$; and where the base, x, is an integer greater than or equal to 2. The number has the value

$$n = s(a_k \times x^k + a_{k-1} \times x^{k-1} + \ldots + a_1 \times x^1 + a_0 \times x^0).'$$

The expression for n is called the *literal* expansion of the number n.

The integer zero is represented by '0' in all bases. It is neither positive nor negative so has no sign. If s is '$+$' it is usual to omit the sign.

The phrase 'an integer is represented as a string of length n in base x' means 'the integer is written in base x, with sufficient leading zeros attached to make a string of n digits'. If the base is 2, the string may be referred to as a *binary* string. If the base is 10, the string may be referred to as a *decimal* string.

Example 1:   Give the literal expansion of
$325(base10), -1101(base2), 1703(base8)$.

$$
\begin{aligned}
325(\text{base } 10) &= 3 \times 10^2 + 2 \times 10^1 + 5 \times 10^0 \\
-1101(\text{base } 2) &= -(1 \times 2^3 + 1 \times 2^2 + 0 \times 2^1 + 1 \times 2^0) \\
1703(\text{base } 8) &= 1 \times 8^3 + 7 \times 8^2 + 0 \times 8^1 + 3 \times 8^0
\end{aligned}
$$

Hexadecimal requires the invention of six new digits. The decimal numbers 10, 11, 12, 13, 14, 15 are represented by A, B, C, D, E, F respectively.

Example 2:   Give the literal expansion of A13B(base 16).

$$A13B(\text{base } 16) \quad = \quad 10 \times 16^3 + 1 \times 16^2 + 3 \times 16 + 11$$

The notation $325_{10}, 1101_2, 1703_8, A13B_{16}$ is also used to indicate the base.

Example 3:   Represent 325_{10} as a decimal string of length 8.

$$00000325$$

Example 4:   Represent 1101_2 as a binary string of length 8.

$$00001101$$

2.1.1 Changing base

In changing base the sign of an integer is not affected so it will be sufficient to give the rules for positive integers.

Changing from decimal

Problem 2.1 *How can the representation of an integer be changed from base 10 (decimal) to another base?*

The method involves repeated integer division of the number by the base. The method is illustrated by the following examples.

Example 1: Convert 380_{10} to octal.

In the calculation, successive quotients are divided by the base until the quotient is reduced to zero. The remainders are recorded, right to left, as octal digits.

$$
\begin{aligned}
380 &= 8 \times 47 + 4 \quad \text{(divide by 8, record 4)} \\
47 &= 8 \times 5 + 7 \quad \text{(divide by 8, record 7)} \\
5 &= 8 \times 0 + 5 \quad \text{(divide by 8, record 5)}
\end{aligned}
$$

The result is $380_{10} = 574_8$. The result may be verified by substituting successive quotients back into the line above, thus:

$$
\begin{aligned}
380 &= 8 \times (8 \times (8 \times 0 + 5) + 7) + 4 \quad \text{(decimal)} \\
&= 5 \times 8^2 + 7 \times 8^1 + 4 \times 8^0 \quad \text{(decimal)} \\
&= 574_8
\end{aligned}
$$

Example 2: Convert 380_{10} to hexadecimal.

Successive quotients are divided by the base until the quotient is reduced to zero. The remainders are recorded, right to left, as hexadecimal digits.

$$
\begin{aligned}
380 &= 16 \times 23 + 12 \quad \text{(divide by 16, record C)} \\
23 &= 16 \times 1 + 7 \quad \text{(divide by 16, record 7)} \\
1 &= 16 \times 0 + 1 \quad \text{(divide by 16, record 1)}
\end{aligned}
$$

The result is $380 = (1)(7)(12) = 17C(\text{base } 16)$. The result may be verified by substituting successive quotients back into the line above, thus:

$$
\begin{aligned}
380 &= 16 \times (16 \times (16 \times 0 + 1) + 7) + 12 \quad \text{(decimal)} \\
&= 1 \times 16^2 + 7 \times 16^1 + 12 \quad \text{(decimal)} \\
&= 17C(\text{base } 16)
\end{aligned}
$$

The algorithm below describes the method in more general fashion.

Algorithm 2.1 *To change the representation of an integer from decimal to another base.*

input:	An integer, $n > 0$, represented in decimal.
	An integer $b \geq 2$. *(b is the new base)*
	Precondition: $n > 0$, n represented in decimal, $b \geq 2$.
output:	n, represented in base b.
method:	Let q have initial value n.
	While $q > 0$
	divide q by b to give remainder r, quotient q_1
	let the new value of q be the quotient so found, (q_1)
	let the next digit(right to left) in the output be r.

Exercise: Use the algorithm to show $380_{10} = 101111100_2$.

Changing to decimal

Problem 2.2 *How can the representation of an integer be changed to decimal from another base?*

The problem is just the reverse of the first problem. The most obvious procedure would be to multiply out the literal expansion of the integer. However the method below is more efficient, requiring fewer operations. It depends on rebracketing the literal expansion.

Example: Convert 23DA(base 16) to decimal.

From the definition:

$$
\begin{aligned}
23\mathrm{DA}(\text{base } 16) \;=\;& 2 \times 16^3 + 3 \times 16^2 + 13 \times 16 + 10 \\
& \textit{(6 multiplications and 3 additions)} \\
=\;& 16 \times (2 \times 16^2 + 3 \times 16 + 13) + 10 \\
=\;& 16 \times (16 \times (2 \times 16 + 3) + 13) + 10 \\
=\;& 16 \times (16 \times (16 \times 2 + 3) + 13) + 10
\end{aligned}
$$

(Calculation of the last expression requires 3 multiplications and 3 additions.)

The calculation proceeds from the inner bracket to obtain 9178 in decimal.

The algorithm used in the above example is usually called Horner's method.

Algorithm 2.2 (Horner) *To change the representation of an integer to decimal from another base.*

> **input:** x,   *(the base in which the number n is represented)*
> $a_k, a_{k-1}, ..., a_0$.   *(the digits of n in base x)*
>   Precondition: $x \geq 2$ and $a_k \neq 0$ and $0 \leq a_j < x$.
> **output:** n, the number expressed in base 10.
> **method:**   Let n have initial value a_k.
>   For $i = 1$ to k
>   replace n by $n \times x + a_{k-i}$.
>   *(i.e. n times the base plus the next digit to the right)*

The following examples illustrate the application of the algorithm.

Example 1:   Convert $23DA_{16}$ to decimal.

Representing numbers in decimal within the steps of the algorithm, we have:

x	a_3	a_2	a_1	a_0		n
16	2	3	13	10		2
					$2 \times 16 + 3 = 35$	35
					$35 \times 16 + 13 = 573$	573
					$573 \times 16 + 10 = 9178$	9178

hence $23DA_{16} = 9178_{10}$

Example 2:   Convert 7025_8 to decimal.

x	a_3	a_2	a_1	a_0		n
8	7	0	2	5		7
					$7 \times 8 + 0 = 56$	56
					$56 \times 8 + 2 = 450$	450
					$450 \times 8 + 5 = 3605$	3605

hence $7025_8 = 3605_{10}$

2.1.2   Arithmetic operations on positive integers

The ordinary algorithms that we use for arithmetic operations on positive integers depend on the base, that is on 10. It is possible to construct a general algorithm for any base for each operation. The algorithms could be extended to deal with negative integers as well as positive integers by adding the appropriate rules. However the algorithms given below cover the interesting part, dealing with the 'carry digits'.

Problem 2.3 (Addition) *How may we add two positive integers repre-sented in base x?*

Let us carry out an addition in decimal, noting the part played by the base, and then carry out parallel calculations in a different base.

Example: Compute $27 + 15$ in decimal, binary and octal.

	decimal	*binary*	*octal*
'carry digit'	10	111110	10
	27	11011	33
	15	1111	17
	42	101010	52

In each case the answer is forty-two.

The algorithm below generalises the method to an arbitrary base.

The symbol '←' used in the algorithm below, assigns to the variable on the left the value of the expression on the right. Some writers use the symbol ':=', borrowed from the computer language *Pascal*, to convey the same meaning.

For example, the line $c_i \leftarrow a_i + b_i + d_i - x$ means that the value of c_i is to be replaced by the value of $a_i + b_i + d_i - x$.

Algorithm 2.3 (Addition) *To add two positive integers represented in base x.*

> **input:** a, b, integers, each represented in base x.
> A precondition is $a > 0, b > 0$.
>
> **output:** c, integer, represented in base x. Postcondition $c = a + b$.
>
> **method:**   Represent a by $a_k \cdots a_0$, a string of length $k + 1$ in base x.
> Represent b by $b_k \cdots b_0$, a string of length $k + 1$ in base x.
> (if necessary, fill up with leading zeros so that the strings are equal)
> $d_0 \leftarrow 0$ *(d_i is to be the 'carry digit' from column $i - 1$)*
> For $i = 0$ to k
> $$\begin{cases} \text{if } a_i + b_i + d_i \geq x \\ \qquad \text{then } \begin{cases} d_{i+1} \leftarrow 1 \\ c_i \leftarrow a_i + b_i + d_i - x \end{cases} \\ \qquad \text{else } \begin{cases} d_{i+1} \leftarrow 0 \\ c_i \leftarrow a_i + b_i + d_i \end{cases} \end{cases}$$
> if $d_{k+1} = 1$ then $c_{i+1} \leftarrow 1$

Problem 2.4 (Subtraction) *How may we subtract one integer from another, each represented in base x?*

The decimal subtraction below reminds us of the part played by the base(10) in ordinary subtraction. The binary and octal subtractions illustrate parallel calculations in different bases.

Example: Compute $50 - 22$.

decimal	*binary*	*octal*
50	110010	62
-22	-10110	-26
28	11100	34

In each case the answer is twenty-eight.

Algorithm 2.4 (Subtraction) *To subtract one positive integer from another, each represented in base x.*

    input: a, b, integers, each represented in base x.
                Represent a by $a_k \cdots a_0$, a string of length $k + 1$.
                Represent b by $b_k \cdots b_0$, a string of length $k + 1$.
                (if necessary, fill up with leading zeros
                so that the strings are equal)
                A precondition is $(a > b > 0)$.
    output: c, integer, represented in base x. Postcondition $c = a - b$.
    method: $d_0 \leftarrow 0$ (*d_i is to be the 'carry digit' from column $i - 1$*)
                For $i = 0$ to k

$$\begin{cases} \text{if } a_i - b_i - d_i \geq 0 \\ \qquad \text{then} \begin{cases} d_{i+1} \leftarrow 0 \\ c_i \leftarrow a_i - b_i - d_i \end{cases} \\ \qquad \text{else} \begin{cases} d_{i+1} \leftarrow 1 \\ c_i \leftarrow a_i + x - b_i - d_i \end{cases} \end{cases}$$

Multiplication

An algorithm for the multiplication of positive integers represented in an arbitrary base x could use repeated addition. The ordinary algorithm for multiplication relies on knowledge of a 'multiplication table' for the base in which the integers are represented and uses the fact that, for each base x, multiplication by x^n 'shifts' the integer representation n places to the left by placing n zeros to the right. The multiplication table for binary consists of the following four facts:

$$0 \times 0 = 0 \quad 0 \times 1 = 0$$

$$1 \times 0 = 0 \quad 1 \times 1 = 1$$

The following examples illustrate the ordinary algorithm for multiplication of integers represented in base 10 and the parallel versions for base 2 and base 8.

Example: Compute 27×13.

decimal	*binary*	*octal*
27	11011	33
$\times$ 13	$\times$ 1101	$\times$ 15
81	11011	207
270	1101100	330
351	11011000	537
	101011111	

Division

An algorithm for the division of positive integers represented in an arbitrary base x could use repeated subtraction. The following examples illustrate the ordinary algorithm for division of integers represented in base 10 and the parallel version for base 2.

Example 1: Compute $729 \div 17$.

$$
\begin{array}{r}
\text{decimal} \\
42 \\
17 \overline{)\ 729} \\
680 \\
\hline
49 \\
34 \\
\hline
15
\end{array}
$$

- 42   quotient
- 680   subtract 40×17 from 729
- 49   remainder so far
- 34   subtract 2×17 from 49
- 15   remainder

The quotient is 42 and the remainder 15.

Example 2: Compute $110110_2 \div 1001_2$ in binary.

```
                      binary
                110          quotient
       1001  )  110110
                100100       subtract 100 × 1001
                 10010       remainder so far
                 10010       subtract 10 × 1001
                     0       remainder
```

The quotient is 110_2 and the remainder is 0.

2.1.3   Representation of integers in a computer

This section is about the representation of integers in a computer. The properties of the integers viewed as a mathematical system are embodied in axioms and it is desirable that these properties be preserved in the computer representation. For most practical arithmetic there is no difficulty, but for 'large' numbers the first axiom fails and there may be problems with some of the others.

Integer axiom 1. For each pair of integers a, b, $a + b$ and ab are unique integers. We say that addition and multiplication are *defined* on the integers or sometimes that the integers are *closed* under addition and under multiplication.

No matter how much space is reserved for each integer there will be some integers too 'big' for accurate representation in a computer. For example the product

$$123123123 \times 987987987$$

cannot be carried out accurately on my calculator. The calculator returns the approximate answer 1.2164417×10^{17}. For positive integers greater than 10^{100} my calculator just returns an error message; so axiom 1 ultimately 'fails'.

Positive integers

For the storage of information, a device that can exist in different states, and that can be switched accurately between those states, is needed. It must be possible to identify the state of the device without error. The type of device that is most reliable for accurate storage and identification is one that has just two states. Therefore, at the basic level, information should be arranged in some two-value form.

For integers the appropriate arrangement is the binary system, which uses only the digits 0 and 1. The elementary device which can exist in two states is called a 'bit'; for our purposes the states are labelled 0 and 1. The word *bit* is a contraction of '*b*inary dig*it*'.

How should the potentially unlimited size of numbers be recorded in a machine that must have limited capacity?

The practice is to decide on the number of bits used to represent an integer. The sequence of bits is called a 'fixed length word' or 'word'. Some commonly used choices are 8-bit, 16-bit and 32-bit words. Let us illustrate the concept using 4-bit words:

0	is represented by	0000
1		0001
2		0010
3		0011
4		0100
⋮		⋮
7		0111

You will probably notice that the 4-bit strings above correspond to the binary numbers from 1 to 7. Since each bit has two possible values there will be 2^4 or 16 possible words in all, leaving eight words to be allocated to negative integers. The allocation has been done in several ways, perhaps the most obvious being the 'sign, magnitude' method which uses one bit for the sign and the rest for the magnitude of the number. However the following is probably the most often used method. It is called the 'two's complement method'. The advantage of the 'two's complement method' is that you just go ahead and add (or subtract) without having to check signs first. An 'overflow' error can occur but this is (electronically) easy to detect.

Negative integers

Continuing with 4-bit words for illustration, we have:

-1	is represented by	1111
-2		1110
-3		1101
-4		1100
⋮		⋮
-7		1001
-8		1000

Notice that words representing negative integers each have a '1' in the first position. But how is the whole word chosen?

The 4-bit (fixed length) word 1111, regarded as a (variable length) binary integer, is equal to $10000 - 1$.

Thus, when we find $^-1$ is represented by 1111 in the table above, this corresponds to $10000 - 1$ in binary, that is $16 - 1$ in decimal. Similarly $^-2$ is represented by 1110 which corresponds to $10000 - 10$ in binary or $16 - 2$ in decimal. The explanation is that the 4-bit words correspond to numbers in $\mathbf{Z}_{16}$ in which $16 \equiv 0$.

Let us do the binary calculation that produces 1110 for $^-2$.

$$
\begin{array}{rl}
10000 & \text{may be written} \\
-\quad 10 & \\
\hline
1110 & \text{which gives} \\
& \text{or}
\end{array}
\qquad
\begin{array}{rl}
1111 & +1 \\
-\quad 10 & \\
\hline
1101 & +1 \\
\hline
1110 &
\end{array}
$$

The calculation on the left uses more than four bits. The advantage of the calculation on the right is that it can be done correctly using no more than four bits.

Observe that 1101 has digits which are the 'complements' of the digits in 0010, the binary word for 2. This leads to the following procedure for finding the fixed length word representation for $^-2$. It is a procedure that can easily be automated in computer hardware.

1. Write down the word for 2, giving 0010.

2. Replace each 0 by 1 and each 1 by 0, giving 1101.

3. Add 1 to 1101 in binary, giving 1110. (For the last step to apply to all cases, ignore any carry digit to column 5.)

The reason for the statement in parentheses in step 3, is to deal correctly with the negative of 0. There is only one integer whose negative is itself and that integer is 0. Apply the above procedure to finding the fixed length word representation for $^-0$.

1. The word for 0 is 0000.

2. The transformation yields 1111.

3. Adding 1 to 1111 gives 0000 (the mathematically correct result) provided we ignore the last carry digit.

The procedure can be used to find the representation for the negative of any integer, provided that we do not need to go outside the available words. For example, the negative of $^-8$ is not available in 4-bit words.

Representing integers with fixed 4-bit words is somewhat limited. A computer using 16-bit words, such as an IBM PC has 2^{16}, that is 65536 possible words. Using the two's complement scheme, the integers from -32768 to 32767 may be represented. Many computers use 32-bit words, which allow representation from -2147483648 to 2147483647.

We realise that axiom 1 for integers does not hold for their computer representation. Nevertheless quite a lot of arithmetic is possible within the ranges used in practice. All of the axioms for integers hold, provided calculations remain within the allowed range. If a calculation results in an answer that cannot be represented within the fixed length words allowed, we say that 'overflow' has occurred. If overflow occurs in a computer a signal, warning of this fact, is generated.

Now consider fixed length words of arbitrary length.

Problem 2.5 *How may we represent positive integers, using fixed length n-bit words?*

Algorithm 2.5 *To represent a non-negative integer, i, using n-bit words.*

input:	n, integer
	(the fixed number of bits in an integer word)
	i, integer. Precondition: $0 \le i \le 2^{n-1} - 1$.
	(i is the integer to be represented)
output:	An n-bit word representing i.
method:	(if necessary) Represent i in binary
	$a_k a_{k-1} \cdots a_0$, where $k \le n - 2$
	Fill up with leading zeros to make an n−bit word.

Example: Represent 53 in 8-bit words.

$$53_{10} = 110101_2.$$

So 53 is represented by 00110101.

Problem 2.6 *How may we represent the negative of an integer using n-bit words?*

Algorithm 2.6 (Negation) *To represent the negative of an integer using n-bit words. (The two's complement representation of signed integers.)*

input: n, integer.

i, integer. Precondition: $-2^{n-1} \le i \le 2^{n-1} - 1$

output: An n-bit word that represents ^-i.

method: Case 1. $i \ge 0$.

1. If necessary, use algorithm 2.5 to write the n-bit word for i.

2. Replace digits. Replace 0 by 1 and 1 by 0.

3. Add 1 to the n-bit word so obtained, using binary addition except that a 'carry' digit from column n should be ignored.

4. The word so produced represents ^-i.

Case 2. $i < 0$.

If the representation for i is given, apply steps 2,3,4 of case 1; otherwise $^-i > 0$ and ^-i can be represented directly using algorithm 2.5.

Example 1: Represent the negative of 53 using 8-bit words.

$$53_{10} \quad \text{is represented by}$$

	00110101	
	11001010	*(replacing digits)*
+	1	*(adding 1)*
	11001011	

$(^-53)$ is represented by 11001011

Example 2: Represent $^-(^-53)$ using 8-bit words.

$$(^-53)_{10} \quad \text{is represented by}$$

	11001011	from example 1
	00110100	*(replacing digits)*
+	1	*(adding 1)*
	00110101	

$^-(^-53)$ is represented by 00110101

Example 3: Find the integer whose 8-bit word is 11100001. The required integer must be negative, since the leftmost digit is 1. We use the negation algorithm to discover its value. Let the integer whose representation is 11100001 be i. Then:

i is represented by   11100001
apply algorithm 2.6
replace digits: 00011110
add 1:   00011111
$\bar{\ }i$ is represented by   00011111
so $\bar{\ }i$ equals   31
and $\bar{\ }(\bar{\ }i)$ equals $\bar{\ }$31

So by theorem 2.1(a)

$$i = \bar{\ }31.$$

The method by which fixed length n-bit words are 'added' is to treat them as binary integers, except that carry digits from the leftmost column are ignored.

2.2   REAL NUMBERS

The operation of division is not defined on the integers. That is to say $a \div b$ is not an integer for all pairs of integers a, b.

However, the integers may be extended to form a new number system, called the *rationals*, in which division *is* defined.

Even the rationals are not sufficient to do all that mathematicians would like. To solve equations such as $x^2 - 2 = 0$ and to represent every point on a line by a number, it is necessary to invent another class of number called the *irrationals*.

The rational numbers together with the irrational numbers form the *real* numbers.

2.2.1   Rational numbers

Rational numbers are basically integer pairs in a definite order. For example $\frac{2}{3}$, sometimes written 2/3, is a rational number, and 3/2 is a different one. The set of integers is denoted by **Z**. The set of rational numbers consists of all ordered pairs of integers p/q in which $q \neq 0$. It is usually denoted by **Q**. The four operations, addition, multiplication, subtraction and division, are defined on **Q**.

For all rationals $\frac{p}{q}, \frac{r}{s}$, where p, q, r, s are integers and $q, s \neq 0$, we define:

addition

$$\frac{p}{q} + \frac{r}{s} = \frac{ps + qr}{qs}$$

subtraction

$$\frac{p}{q} - \frac{r}{s} = \frac{ps - qr}{qs}$$

Note that this definition is equivalent to definition 2.1 for subtraction if we define the negative of the rational $\frac{r}{s}$ to be $\frac{^-r}{s}$.

multiplication

$$\frac{p}{q} \times \frac{r}{s} = \frac{pr}{qs}$$

division provided $r \neq 0$

$$\frac{p}{q} \div \frac{r}{s} = \frac{ps}{qr}$$

Note that this definition may be rephrased in terms of the definition of multiplication, if we define the inverse of a rational $\left(\frac{r}{s}\right)$ to be $\frac{s}{r}$.

Then, to divide by a rational, multiply by its inverse.

But the rationals introduce a new problem. A rational number may be represented by infinitely many different integer pairs so we need a mechanism to decide whether distinct integer pairs represent the same rational or not.

For example 1808/1017 and 16/9 are the same number.

One approach to this problem is to define equality of rationals.

Definition 2.1 (equality) $\frac{p}{q} = \frac{r}{s}$ *if and only if* $ps = qr$.

Another approach is to express a rational $\frac{p}{q}$ in a standard form.

This may be done by expressing the rational in 'lowest terms', for example:

$$\frac{1808}{1017} = \frac{113 \times 16}{113 \times 9}$$

$$= \frac{16}{9}$$

where 113 was found as the gcd of 1808 and 1017. Or it may be done by extending the 'literal expansion' of a number to rationals, by the usual process of division, extended beyond the decimal point, e.g.

$$1808/1017 = 1.7777\cdots$$
$$= 1.\dot{7}$$
$$16/9 = 1.7777\cdots$$
$$= 1.\dot{7}$$

The last alternative introduces another difficulty. My calculator returns 1.7777778 for 16/9. You may feel this is close enough; nevertheless, to cope with the finite size of storage space, it has been necessary to approximate the correct answer. The study of calculations that are sensitive to small errors is part of numerical analysis, which is too specialised for us to pursue here.

There are two issues for us to consider:

- How should a number, whose literal expansion is such that its representation would overflow the space reserved for numbers, be represented on a machine?

- How should the algorithm for determining the decimal representation of a rational number be terminated?

The first question may be investigated by performing the following on your calculator:

$$77777777 + 0.12345678 - 77777777$$

My calculator returns 0.12; others produce different answers. The answer shows how many places are held by the calculator for the purpose of *rounding* the visible number to the nearest value to what is correct. For my calculator, if the hidden places are in the range 00-49, then no change is made to the visible number. If the hidden places are in the range 50-99, then the last visible digit is increased by 1.

The second question may be investigated by doing some examples.

Example 1: Find the literal expansion of the rational 8/11 in decimal.

$$
\begin{array}{r}
0.72 \\
11 \;)\; \overline{8.00} \\
\underline{7.70} \\
30 \\
\underline{22} \\
8
\end{array}
$$

At this point the algorithm requires the operation '8 divided by 11', which is the same as the first operation. The algorithm repeats the same sequence of divisions as began the process. The algorithm will cycle endlessly through the digits 7, 2. We have:

$$8/11 = 0.7272\cdots = 0.\dot{7}\dot{2}$$

Example 2: Find the literal expansion of the rational 3/4 in decimal.

For this example the same procedure of division terminates. We have:

$$3/4 = 0.75$$

Example 3: Find the literal expansion of the rational 13/61 in decimal.

Using the division algorithm as in the above examples we obtain:

13/61 = 0.21311475409836065573770491803278688524590163934426 2295081967, after which the digits repeat.

For division by 61 there are 60 possible remainders other than 0, so the largest possible cycle is 60; and this is achieved for 13/61.

The literal expansion of p/q where p, q are integers, either terminates or cycles with cycle length at most $q - 1$. The simplest rule is to stop when some specified number of digits has been reached.

The algorithm used on my calculator appears to stop after 9 digits have been calculated and then the last two digits are used to round the visible number correct to 7 figures. My calculator would yield

$$0.7272727, \quad 0.75, \quad 0.2131148$$

to the examples given above.

Rational numbers expressed in binary

The string of digits

$$a_k \cdots a_1 a_0 . a_{-1} \cdots a_{-s} \quad (\text{base } x)$$

represents the literal expansion

$$a_k x^k + \cdots + a_1 x + a_0 + a_{-1} x^{-1} + \cdots a_{-s} x^{-s}$$

Example 1: Convert 11.011_2 to decimal.

$11.011_2 = 2 + 1 + \frac{1}{4} + \frac{1}{8} = 27/8 = 3.375$ (decimal) by everyday division (or by calculator).

Example 2: Convert 1101.101101_2 to decimal.

By Horner's algorithm (2.2) we have $1101101101_2 = 877_{10}$. Thus:

$$
\begin{aligned}
1101.101101 &= 1101101101/1000000 \quad \textit{(binary)} \\
&= 877/64 \quad \textit{(decimal)} \\
&= 13.703125 \quad \textit{(decimal)}
\end{aligned}
$$

The above examples show two approaches to the conversion of a rational expressed in binary to decimal.

Problem 2.7 *How can we convert a rational fraction expressed in decimal to binary?*

The approach used below is to subtract out the largest binary fraction possible at each stage, until nothing is left.

For example, $0.75 - \frac{1}{2} - \frac{1}{4} = 0$. Thus $0.75_{10} = 0.11_2$.

Many will prefer the algorithm of question 2.4.18, because it is easy to apply and has the advantage that 'cycling' may be identified. The algorithms may be compared by applying them to the same number, 0.6 say.

Algorithm 2.7 *To convert a rational fraction from decimal to binary.*

    input: Rational number a, between 0 and 1, expressed in decimal.

    output: a expressed in binary correct to eight digits (say). The digits are recorded from the point to the right, in order as found.

    method: Let a be the rational number to be converted to binary.
                  Give i initial value 1.   *(i counts the number of places*
                                      from the point)
                  While $a > 0$ and $i \leq 8$:
                      if $2^{-i} \leq a$
                          then
                                a is given the value $a - 2^{-i}$
                                record 1
                    else
                          record 0
                increment i by 1

    Example: Convert 0.375_{10} to base 2.

i	a	2^{-i}	$(2^{-i} \leq a)$	*output*
1	0.375	0.5	F	.0
2	0.375	0.25	T	.01
3	0.125	0.125	T	.011
4	0			

Errors

Let $y = x + e$ be an approximation for x. Then $|e|$ is called the absolute error in taking y as an approximation for x. Define ε to be the greatest possible value of $|e|$.

Example: A calculator returns the value 0.714285 for some unspecified calculation. What is the largest possible error: (a) if the calculator 'rounds'? (b) if the calculator 'truncates'?

(a) Six digits are visible after the decimal point. On the assumption that the calculator has further digits hidden and the visible number has been obtained by *rounding*, the last visible digit should be the six digit figure nearest to the correct answer. In practice this means that if the hidden digits represent half or more of the value of the last visible digit, then that digit has been increased by one. Numbers in the range $0.7142845 - 0.7142854999 \cdots$ round to the given answer. The greatest possible absolute error is $\varepsilon = 0.0000005$.

(b) Truncation is the action of 'cutting off' or ignoring the hidden digits. Numbers in the range $0.714285 - 0.714285999 \cdots$ truncate to the given answer. The greatest possible absolute error is $\varepsilon = 0.000001$.

absolute error

$$|e| = |y - x|$$

relative error

$$\frac{|e|}{|x|} = \frac{|y - x|}{|x|}$$

relative error percent

$$\frac{100|e|}{|x|} = \frac{100|y - x|}{x}$$

In practice we would divide by $|y|$, since x would be unknown, and replace e by ε.

2.2.2 Irrational numbers

Every rational number ($\frac{p}{q}$) has a decimal expansion that either terminates or cycles endlessly through the same sequence of digits. But there are numbers with decimal expansions that neither terminate nor cycle. Such numbers cannot be expressed as the ratio of two integers and so are called *irrational*.

Examples of irrational numbers are $1.1010010001 \cdots$, $\sqrt{2}$, π, $\log_{10} 7$, $\ln 3$, $\sin \frac{\pi}{10}$. There are infinitely many irrationals.

For storage in a machine, irrational numbers are approximated by fixed length rationals, in the same way as rationals that cycle or are just too long.

An iterative algorithm for calculating $\sqrt{k}$

Problem 2.8 *How can we find the square root of a positive real number?*

The following is an efficient algorithm for a calculator or computer to find the square root of a positive number.

Algorithm 2.8 *To calculate to some given accuracy the positive square root of a positive real number.*

> **input:** k, real; $k > 0$.
> *(we wish to find the square root of k)*
> ε, real; $\varepsilon > 0$.
> *(ε is the measure of 'closeness' that will be accepted)*
>
> **output:** y, real.
> *(y is the final approximation to $\sqrt{k}$).*
>
> **method:** Give x a value; any positive number will do.
> $i \leftarrow 0$ *(give i the initial value 0)*
> Repeat
> $$\begin{cases} x &\leftarrow& (x + k/x)/2 \\ & & \text{(}x\text{ is the current approximation to } \sqrt{k}\text{)} \\ i &\leftarrow& i+1 \end{cases}$$
> until x is close to $\sqrt{k}$.
> $y \leftarrow x$

What does *close* mean?
x is close to $\sqrt{k}$ may be refined to

$$(|x^2 - k| \le \varepsilon) \text{ is true} \tag{1}$$

$$\text{or } (|x - \text{ the previous } x| \le \varepsilon) \text{ is true} \tag{2}$$

Statements (1) and (2) are Booleans. The first statement is true if the square of x is close to k. The second statement is true if successive values of x are close to each other.

Omitting the variable i would have no effect on the operation of the algorithm. It merely acts as a counter to give the number of the current approximation (x) to $\sqrt{k}$.

Example 1: Compute $\sqrt{25}$.

Give x initial value $25/2$. Set ε to 0.02. Use 'closeness' criterion (2).

x	i	$|x_i - x_{i-1}|$	$(|x_i - x_{i-1}| \le 0.02)$
12.5	0		
7.25	1	5.25	F
5.349	2	1.9	F
5.011	3	0.338	F
5.000013	4	0.011	T

Example 2: Compute $\sqrt{10}$. What is the percentage relative error in taking $\sqrt{10}$ as an approximation for π?

Give x initial value 5. Set ε to 0.0001. Use 'closeness' criterion (1).

x	i	x^2	$x^2 - 10 \leq 0.0001$
5	0		
3.5	1	12.25	F
3.17857	2	10.1033	F
3.162319	3	10.00026	F
3.162277	4	10.00000	T

Correct to four decimal places $\sqrt{10} = 3.1623$; but $\pi = 3.1415926\cdots$

The percentage relative error is

$$|3.1623 - \pi| \times 100/\pi = 0.7\%$$

Is the algorithm correct? It can be shown using inequalities that $\sqrt{k}$ lies between x and k/x; and that averaging these two values gives a better approximation to the square root of x.

A better (i.e. simpler and more general) proof that the algorithm is correct uses Newton's method, but that method uses calculus, which is outside the range of this book.

2.3 EXERCISES

1. Give the literal expansion of:

   (a) 1209_{10} (b) 110101_2 (c) $C10AA_{16}$

2. Convert the following numbers to binary, then to hexadecimal:

   (a) 234_{10} (b) 1024_{10} (c) 255_{10} (d) 747_8 (e) 326_8

   Explain, in ordinary English, how to do (d) and (e) without converting to decimal.

3. Convert to decimal:

   (a) ABCD(hexadecimal)   (b) 12073(octal)   (c) 1011101(binary)

4. Calculate the following in binary:

   (a) $10110 + 1001$ (b) $101010 - 11011$ (c) $110 * 110$ (d) $1110100/101$

5. Calculate the following in hexadecimal:

(a) 347A + 8741   (b) 23B6 − 428

6. Walk through algorithm 2.3 given $x = 8, a = 2304, b = 514$. Give the output from the algorithm.

7. Walk through algorithm 2.4 given $x = 2, a = 1001, b = 101$. Give the output from the algorithm.

8. Give the twos-complement representation of 63 and of ⁻17 in 8-bit words.

9. Find the integers represented in twos-complement form by the 8-bit words 00111001, 10011010.

10. Design an algorithm to:

(a) solve $ax = b$ for x in terms of the real numbers a, b.

(b) solve $ax^2 + bx + c = 0$ for x in terms of the real numbers a, b, c.

11. A Babylonian clay tablet of about 1700 BC gives the value 1;24,51,10 for the square root of 2. Given that the ancient Babylonians used a sexagesimal system of numeration, that is a system based on 60, calculate their value for $\sqrt{2}$ and determine its accuracy.

12. Which, if either, of

$$\frac{215491}{216169} \quad \text{or} \quad \frac{215492}{216170} \quad \text{equals} \quad \frac{1907}{1913}?$$

13. Convert $x = 14.14$ from decimal to binary.
Round the answer to a binary number y with 10 binary digits. Find the relative error of y as an approximation to x by converting y back to decimal form.

14. The rational number $\frac{22}{7}$ is often used as an approximation to π. Given that $\pi = 3.141592654\cdots$, find the relative error in the given approximation.

15. Calculate the square root of 5 using algorithm 2.8.

What is the absolute error if two iterations are performed, starting with $x = 2$ as the first approximation to $\sqrt{5}$?

16. Archimedes (287-212 BC) states $\dfrac{1351}{780} > \sqrt{3} > \dfrac{265}{153}$. Which of these approximations is closer to $\sqrt{3}$?

What is the absolute error in taking $\frac{1351}{780}$ as an approximation to $\sqrt{3}$?

17. Convert: (a) 2324.05(base 10) to base 16

            (b) 100010.1111(base 2) to base 10

            (c) 1AA.2(base 16) to base 2

18. Consider the following algorithm:

> **input:** x, rational, between 0 and 1, expressed in decimal.
>
> **output:** x expressed in binary correct to eight digits.
>
>   Record digits as found, from the point to the right.
>
> **method:** $a \leftarrow x$
>
> $i \leftarrow 1$
>
>   While $a > 0$ and $i \leq 8$:

$$
\left\{
\begin{array}{l}
a \leftarrow a * 2 \\[4pt]
\text{if } a \geq 1 \\[4pt]
\qquad \text{then} \\[4pt]
\qquad\qquad \left\{ \begin{array}{l} a \leftarrow a - 1 \\ \text{record } 1 \end{array} \right. \\[12pt]
\qquad \text{else} \\[4pt]
\qquad\qquad \text{record } 0 \\[4pt]
i \leftarrow i + 1
\end{array}
\right.
$$

(a) Given the input $x = 0.6_{10}$, find the output.

(b) Explain in words what the algorithm does. How would you identify 'cycling' in the output?

19. The algorithm given below finds positive integer powers of a real number.

> **input:** x, real; n, positive integer.
>
> **output:** p, real. (p *should have the value* x^n)
>
> **method:**   Let p have the value 1.
>
>   While $n \neq 0$:
>
>   divide n by 2 to obtain quotient q, remainder r
>
>   if $r = 1$
>
>   then replace p by $p * x$
>
>   let n have the value q
>
>   let x have the value $x * x$

Step through the algorithm for input $x = 3$ and $n = 13$.

20. The following is an iterative algorithm to calculate $\sqrt{2}$.

> **input:** x, real. *(x may be taken as any guess for $\sqrt{2}$)*
> ε, real.
>   Precondition $x > 0;\ \varepsilon > 0$.
> **output:** y, real. *(y is the desired approximation to $\sqrt{2}$)*
> **method:** $a \leftarrow x$
>   While $|a^2 - 2| > \varepsilon$:
> $$\begin{cases} a & \leftarrow & \dfrac{1}{a+1} \\ a & \leftarrow & a+1 \end{cases}$$
> $y \leftarrow a$

(a) Given $\varepsilon = 0.01$ and $x = 2$, step through the algorithm.

(b) What is the absolute error in the approximation for $\sqrt{2}$ found by the algorithm?

(c) What is the purpose of the ε in the algorithm?

Chapter 3

Sets, Functions and Relations

This chapter is about sets, functions, relations and sequences, basic concepts that occur in many branches of mathematics. You will be familiar with at least some of what follows, but we need to set down exactly what is meant by these fundamental ideas. Your aim should be to develop fluency in the accurate use of mathematical terms and in translation between mathematical and ordinary language.

One of our aims is to develop a precise language, based on mathematics, for the specification of algorithms. It is hoped that this will help in the construction of correct programs that can be efficiently maintained, or adapted to meet new requirements. The mathematical language introduced in this chapter is essential for an understanding of the mathematical ideas that follow and for the precise statements we wish to make about programs.

3.1 SETS

Putting things of interest to us into different classes or collections helps to organise our knowledge. After classification, relations between these objects of interest usually emerge to add to our understanding. In a practical situation it should be clear what things are of interest. The totality of objects of interest is called the *universal set*, which is denoted by U. A *set*, S, consists of a well-defined collection of objects from the universal set. 'Well-defined' means that for each object in the universal set, the object definitely belongs to S or the object does not belong to S.

The objects in a set are called the *elements* of the set. The fundamental relation between sets and their elements is *belonging*.

If an element x belongs to a set S, we write $x \in S$.

If on the other hand x does not belong to S we write $x \notin S$.

The usual practice is to use lower case $x, y, a, b...$ for elements and upper case $A, B, S, ...$ for sets. Some frequently used sets may have special notations, for example:

N is the set of natural numbers, $\{0, 1, 2, 3, \cdots\}$, (some writers do not include 0).

Z is the set of integers, $\{\cdots, -2, -1, 0, 1, 2, \cdots\}$.

Q is the set of rationals.

R is the set of reals.

B is the set of Boolean values. $\mathbf{B} = \{\mathsf{T}, \mathsf{F}\}$.

Char is the set of all available characters.

Σ is an 'alphabet' of symbols.

The set with no elements is called the *empty set*, denoted by $\emptyset$ or $\{\ \ \}$.

Example: Let the universal set be *Char*, the set of all characters. Find S, the set of characters in the word 'regret'.

Solution: $S = \{\text{r,g,e,t}\}$.

Notice that the elements may be in any order and they are not repeated. If for some reason you wish to record all repetitions of an element, the corresponding structure is called a *multiset*. For example the multiset of prime factors of 12 is $\{2,2,3\}$.

Example: The Department of Social Security is required to provide a family allowance to families with child dependants if the combined family income is less than a certain figure. The universal set might be all families in the country, and the set to be constructed is all those families eligible for family allowance. For the set to be well-defined, the terms 'family', 'child dependant' and 'combined family income' must be defined without ambiguity. Such definition is not trivial.

A set may be described:

- by listing: $A = \{3, 5, 2, 6, 7, 1, 0, 4\}$.

- by description: A is the set of digits in base 8.

- using a *predicate*:

  $$A = \{x \text{ is an integer } \mid x \text{ is not negative and } x \text{ is less than } 8\}$$

  The symbol '$\mid$' is read 'such that'. The notation '$:$' is also used for 'such that'.

  The expression 'x is not negative and x is less than 8' is true for some integers x and false for the rest. The expression is called a predicate.

A *predicate* must be definitely true or false for each value of the variable (x) in the universe.

The whole thing reads: the set of integers such that the integer is not negative and the integer is less than 8. In symbols the above is written

$$A = \{x \in \mathbf{Z} \mid 0 \le x \le 7\}$$

where $0 \le x \le 7$ is the predicate.

- using set operations such as *union, intersection, difference,...* described below.

- using a characteristic function μ for A.

$$\mu_A(j) = \begin{cases} 1 & \text{if } j \text{ belongs to } A \\ 0 & \text{if } j \text{ does not belong to } A \end{cases}$$

The function μ is defined on all j in the universal set.

Since the universal set, U, contains all the elements of interest for a given problem, the sets of interest will all be subsets of U.

Definition 3.1 (Subset) *The set A is a* subset *of the set B if and only if every element of A is also an element of B. We write $A \subseteq B$.*

The set A is called a proper *subset of B if A is a subset of B and at least one element of B is not in A. We write $A \subset B$.*

Definition 3.2 (Power set) *The collection of all subsets of U is itself a set called the* power *set of U, written*

$$\mathcal{P}(U) = \{X \mid X \subseteq U\}$$

Example: Find the power set of $\{a, b, c\}$.

Each subset of $\{a, b, c\}$ may be constructed using a characteristic function. The eight characteristic functions may be generated from the numbers $0, \cdots, 7$ expressed as three-digit binary strings. For example, 3 yields the binary string 011 and the subset $\{b, c\}$. The table below shows the generation of all the subsets.

binary number	*characteristic*			*subset*
	a	b	c	
1	0	0	1	$\{c\}$
10	0	1	0	$\{b\}$
11	0	1	1	$\{b,c\}$
100	1	0	0	$\{a\}$
101	1	0	1	$\{a,c\}$
110	1	1	0	$\{a,b\}$
111	1	1	1	$\{a,b,c\}$
0	0	0	0	$\{\ \}$

Therefore $\mathcal{P}(\{a,b,c\}) = \{\{c\},\{b\},\{b,c\}\cdots\{\ \}\}$

The power set of $\{a_1, a_2, \cdots a_n\}$ has 2^n elements.

3.1.1 Properties of sets

Having defined subset we can now define equality of sets.

Definition 3.3 (Equality) *Two sets A and B are* equal *if and only if each element in A belongs to B and each element in B belongs to A.*

From the definition of subset above, it follows that the definition of equality is equivalent to

$$A \subseteq B \text{ and } B \subseteq A.$$

Number in a set

The number of elements in a set S is denoted by $|S|$ or $n(S)$. A set containing a single element is called a *singleton*.

3.1.2 Set operations

Each of the following operations produces a new set from two given sets, (A, B).

Intersection

The set of elements that belong to each of two sets in common.

$$A \cap B = \{x \mid x \in A \text{ and } x \in B\}$$

Union

The set of elements that belong to at least one of two sets.

$$A \cup B = \{x \mid x \in A \text{ or } x \in B \text{ or both}\}$$

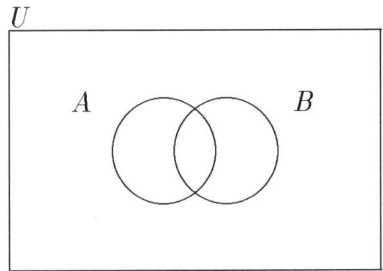

Figure 3.1: Venn diagram for A, B

Difference

The elements of the first set that are not in the second set.

$$A - B = \{x \mid x \in A \text{ and } x \notin B\}$$

Complement

The (set) difference between the *universal* set and the given set. The complement of A is written A'.

$$A' = U - A$$

Symmetric difference

The symmetric difference of A and B, written $A \triangle B$, is the set of elements in exactly one of A, B.

$$A \triangle B = \{x \mid x \in A \text{ or } x \in B \text{ but not both}\}.$$

Example: How many integers from 1 to 1000 are not divisible by 9 or 12?

Solution: Take the universal set to be all integers from 1 to 1000 inclusive.

Let A be the set of integers divisible by 9. Let B be the set of integers divisible by 12.

Then we wish to find $n(U - (A \cup B))$. Figure 3.1 (called a Venn diagram) represents the four possible ways of classifying elements relative to the two sets A, B. That is, in both sets, in neither set, in A alone or in B alone.

$$A = \{9, 18, 27, ...\}, \quad \text{and } n(A) = \lfloor 1000/9 \rfloor = 111$$
$$B = \{12, 24, 36, ...\}, \quad \text{and } n(B) = \lfloor 1000/12 \rfloor = 83$$

From the diagram it is possible to deduce that subtracting 111 and 83 from 1000 has the effect of subtracting the contents of $A \cap B$ twice. Now

$$A \cap B = \{36, 72, ...\} \quad n(A \cap B) = \lfloor 1000/36 \rfloor = 27.$$

$$\begin{aligned} \text{Thus } n(U - A \cup B) &= n(U) - n(A) - n(B) + n(A \cap B) \\ &= 1000 - 111 - 83 + 27 \\ &= 833 \end{aligned}$$

Cartesian product

The set of all *ordered* pairs of elements one from each of two sets, in the order in which the sets are given.

$$A \times B = \{(a, b) \mid a \in A \text{ and } b \in B\}$$

Example 1: $\{3, 4, 2\} \times \{2, 6\} = \{(3, 2), (3, 6), (4, 2), (4, 6), (2, 2), (2, 6)\}$

Notice that $(3,2)$ is present but not $(2,3)$. The order in which the *pairs* appear in the set, however, is unimportant.

Example 2: A screen of text characters can be described by means of sets, using the idea of Cartesian product. Suppose there is a screen that has 40 columns and 20 rows. The usual convention in computing is for columns to increase from left to right and rows to increase going down.

Let C be the set of column numbers: $C = \{c \in \mathbf{N} \mid 1 \le c \le 40\}$.

Let R be the set of row numbers: $R = \{r \in \mathbf{N} \mid 1 \le r \le 20\}$.

Then the set of all screen positions is the Cartesian product $C \times R$.

$$C \times R = \{(c, r) \in \mathbf{N} \times \mathbf{N} \mid 1 \le c \le 40 \text{ and } 1 \le r \le 20\}$$

Notice that this system of labelling points is similar to (but not the same as) the ordinary Cartesian coordinate system.

Let *Char* be the set of characters that can be displayed, including a blank.

A 'screen of text', S, can now be described formally as a set of triples (c, r, x) where x is the character displayed at the location (c, r).

$$S \subset C \times R \times Char$$

Partition

The collection of subsets $A_1, A_2, ..., A_k$ is called a partition of A if and only if every element in A is in one of the subsets A_i, and no element is in more than one of the subsets.

Formally, $A_1, ...A_k$ is a partition of A if and only if

$$A_1 \cup A_2 \cup ... \cup A_k = A$$

and

$$A_i \cap A_j = \emptyset \quad \text{for all pairs } i, j \text{ with } i \neq j.$$

The subsets $A_1, A_2, ...$ are called the *cells* of the partition.

Example 1: Each student in the unit Discrete Mathematics is required to be in one tutorial and only one tutorial. If this rule is followed the tutorials form a partition of the unit Discrete Mathematics.

Example 2: The days of the year each belong to one of seven classes, namely Sunday, Monday, Tuesday, Wednesday, Thursday, Friday, Saturday. For example day 200 in 1991 is a Friday. Each day belongs to one and only one of the days of the week. The days of the week are a partition of the days of the year.

Paradoxes and universal sets

Despite its apparent simplicity the early (pre-axiomatic) set theory led to paradoxes and debate.

Bertrand Russell (1872-1970) used Cantor's (1895) description of a set as 'any collection of definite or separate objects of our intuition or our thought' and attempted to define a set S by

$$S = \{X \mid X \text{ is a set and } X \notin X\}$$

Now some sets are elements of themselves; for example a set of 'thoughts' is a thought, so Russell's 'definition' fits Cantor's description. Russell's question is then 'is S an element of itself?'

If yes, then by the definition $S \notin S$, a contradiction.

If no, then $S \notin S$ so by the definition $S \in S$, a contradiction.

The existence of S is contradictory which is Russell's Paradox. The axiomatic approach to set theory says that the real meaning of the above paradox is that there is no absolute universal set. If we define

$$S = \{X \in U \mid X \text{ is a set and } X \notin X\}$$

where U is a universal set, the paradox disappears. As before, the assertion $S \in S$ leads to $S \notin S$, a contradiction. But the conclusion is now 'S is not in the universe'.

The practical effect of the above discussion is that mathematicians prefer to define a universal set 'constructively' and to use predicates that refer to themselves with care. However we do not wish to reject all definitions that 'refer to themselves'. Such definitions can be very powerful, as in the case of recursive definition.

Fuzzy sets

It may have occurred to you that there are some collections of things that are interesting to us but not well-defined. For example, consider the people that you know who ski. Does that mean regularly, once a year, once a week or what? What level of skill is required before a person can claim to be able to ski? The description 'people that you know who ski' does not define a set.

In 1965, Zadeh introduced the idea of *fuzzy* sets, as a way of dealing with collections whose membership is ill-defined. He defined a 'degree' or 'level' of membership of a 'fuzzy set' on a scale from 0 to 1. This approach might be thought of as an extension of the idea of the characteristic function description of a set.

3.2 FUNCTIONS

A computer program frequently behaves like a function. The *input* to a program may be thought of as the 'x' in a function and the *output* as the unique 'y' associated with x by the function. The 'uniqueness' property is important. It would normally be unacceptable to have different output for the same input on separate executions of a program.

Now for certain inputs a program may not yield an output, for example it may 'loop' indefinitely or halt prematurely. The corresponding situation for a function would be the possibility that for some 'x' values the function may not yield a 'y' value. To allow for the possibility, a *partial function* is defined to include functions that cannot be evaluated for every x. Thus $y = \sqrt{x}$, $x \in \mathbf{R}$ would be a partial function. If the function may be evaluated at every x the function is said to be *total*. A total function has all the properties of a partial function (and so *is* a partial function) with the additional property that for every 'x' there is a 'y'. For example $y = x^2, x \in \mathbf{R}$ is a total function. When, for brevity, the word 'function' is used by itself it will now be taken to mean 'partial function'. A function has three components:

- a set called the *source* of the function.

- a set called the *codomain* or *target* of the function.

- a *rule* that identifies some ordered pairs (x, y), where $x \in source$ and $y \in codomain$, as belonging to the function. For each x the corresponding y is unique.

The *domain* of the function is the set of x's (the first elements of the ordered pairs) and is a subset of the source. The *range* of the function is the set of y's (the second elements of the ordered pairs) and is a subset of the codomain. If the domain of the function is the same set as the source, then the function is *total*.

The above definition does not restrict the source and codomain to sets of numbers.

The 'rule' by which elements (x) in the domain of a function are paired with elements (y) in the codomain has not been defined formally. The rule may be given as an expression for evaluating y for each x or simply as a table of pairs (x, y).

In a first encounter with functions the impression may be gained that the rule *is* the function. For example, some people speak (loosely) of

$$y = 2x^2 + 3x + 5$$

as a function. Certainly $2x^2 + 3x + 5$ is an expression by which y may be evaluated for each x in the domain. But the modern view is that it is the *set of pairs* (x, y) that is the function.

The expression above may be evaluated several ways. For example:

$$x \times (x \times 2 + 3) + 5 \qquad (1)$$
$$\text{or} \quad (2 \times x \times x) + (3 \times x) + 5 \qquad (2)$$

The expressions (1) and (2) represent two distinct algorithms for the evaluation of y at x. The distinction is important to us because the first method of evaluation contains fewer operations than the second. We will look at the efficiency of algorithms in Chapter 7. It seems best now to formally define a partial function and to describe the properties it may have.

Definition 3.4 *A partial function from A to B is a subset f of the Cartesian product $A \times B$, in which for each (x, y) in f, the second element y is unique for each x. That is, if $(x, y_1) \in f$ and $(x, y_2) \in f$ then $y_1 = y_2$.*

The *domain* of f is $\{x \mid (x, y) \in f\}$ and is a subset of A. The set A is called the *source* of the function.

The *range* of f is $\{y \mid (x, y) \in f\}$ and is a subset of B. The set B is called the *codomain* or *target* of f.

A procedure for the evaluation of f at x for each x in the domain of f is called an algorithm for f.

Example 1: The subset of $\mathbf{N} \times \mathbf{N}, f = \{(1,1),(2,1),(3,2)\}$ is a partial function from $\mathbf{N}$ to $\mathbf{N}$. The subset $g = \{(1,1),(1,2)\}$ is not a partial function.

Domain $f = \{1,2,3\}$ and range $f = \{1,2\}$. Each element (x) of the domain is paired with a unique element (y) of the range. The rule for f is just a table of pairs.

If f is a partial function from A to B we write

$$f : A \to B$$

Example 2: A 'greatest common divisor ' partial function may be defined making use of algorithm 1.5 (p. 11). An algorithm may be called by giving its name and providing input values as required in parentheses. It is then understood that the algorithm will return the correct output for the given input. Thus 'algorithm $1.5(a,b)$' where a,b are positive integers returns the positive integer d where $d = \gcd(a,b)$.

The partial function *gcd* is defined by:

$$gcd : \quad \mathbf{Z} \times \mathbf{Z} \to \mathbf{Z}$$

$$gcd(m,n) = \begin{cases} m & \text{if } n = 0, m \neq 0 \\ n & \text{if } m = 0, n \neq 0 \\ \text{algorithm } 1.5(|m|,|n|) & \text{if } m \neq 0, n \neq 0 \end{cases}$$

For this example, domain $gcd = \mathbf{Z} \times \mathbf{Z} - \{(0,0)\}$, which is not equal to the source so the function is not total.

Range $gcd = \{d \in \mathbf{Z} \mid d > 0\}$.

The *image* of f at x is the element (y) paired with x by the function and is written $f(x)$. We write:

$$y = f(x) \quad \text{and say } y \text{ equals } f \text{ of } x \text{ or } f \text{ at } x$$

or

$$f : x \mapsto y \quad \text{and say } f \text{ maps } x \text{ to } y$$

Properties of a function

If the domain of $f = A$ then the function is said to be a *total* function (or mapping).

The function f is said to be *onto B* if the range of f equals B.

The function f is said to be *one-one* when, given any pair x_1, x_2 in the domain of f:

$$\text{if } f(x_1) = f(x_2) \text{ then } x_1 = x_2,$$

or (equivalently)

$$\text{if } x_1 \neq x_2 \text{ then } f(x_1) \neq f(x_2).$$

If the domain of a function is changed then the pairs of the function are changed and by the above definition the function itself is changed.

Example 1: Consider a function *square* whose rule is:

$$square(x) = x^2$$

If the domain is $\mathbf{R}$, then the function is not one-one (counter example: $(-2)^2 = 4$ and $(+2)^2 = 4$).

However, if the domain is restricted to $\mathbf{R}^+$ (the set of positive real numbers) the new function is one-one.

Unfortunately there is some variation in the terminology used by different writers to describe functions and their parts. It is advisable to check definitions if you refer to different books.

Example 2: Identify the domain, source and image of the function:

$$sqrt : \mathbf{N} \rightarrow \mathbf{N}$$

$$sqrt(n^2) = n$$

The domain is the set of perfect squares, $\{0, 1, 4, 9, \cdots\}$. The source is the set of natural numbers $\mathbf{N} = \{0, 1, 2, 3, \cdots\}$. The range is also $\mathbf{N}$. The function is a partial function but not a total function; it is one-one and onto $\mathbf{N}$.

Example 3: Explain why the following function is a total function.

$$sum : \mathbf{N} \times \mathbf{N} \rightarrow \mathbf{N}$$

$$sum(x, y) = x + y$$

sum is a total function since it is defined for all input pairs.

Example 4: Define the absolute value function:

$$abs : \mathbf{R} \rightarrow \mathbf{R}$$

$$abs(x) = |x| = \begin{cases} x & \text{if } x \geq 0 \\ -x & \text{if } x < 0 \end{cases}$$

using set notation.

$$abs : \mathbf{R} \times \mathbf{R} \to \mathbf{R}$$

$$abs = \{(x, y) \in \mathbf{R} \times \mathbf{R} \mid (x \geq 0 \text{ and } y = x) \text{ or } (x < 0 \text{ and } y = -x)\}.$$

The graph of the absolute value function is:

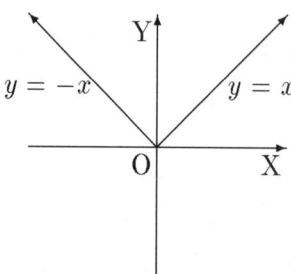

The function *abs* is not one-one.

A counter example that shows the function is not one-one is '$abs(2) = 2$ and $abs(-2) = 2$'. The range is the set of non-negative real numbers, a proper subset of $\mathbf{R}$ so the function is *into* $\mathbf{R}$.

Example 5: $predecessor = \{(n + 1, n) \mid n \in \mathbf{N}\}$

is a partial function from $\mathbf{N}$ to $\mathbf{N}$. It is not a total function since $predecessor(0)$ is undefined.

Example 6: $half = \{(2n, n) \mid n \in \mathbf{N}\}$

is a partial function from $\mathbf{N}$ to $\mathbf{N}$. It is not a total function since, for example, $half(3)$ is not defined.

Example 7: $f : \mathbf{R} \to \mathbf{R}$; $f(x) = 2x + 1$ is one-one and onto $\mathbf{R}$.

For if $2x_1 + 1 = 2x_2 + 1$ then $x_1 = x_2$; and range $f = \mathbf{R}$.

Definition 3.5 (Composition of functions) *Given functions $f : A \to B$ and $g : B \to C$ the composite function $g \circ f$ is defined for all x for which $f(x)$ and $g(f(x))$ exist, by*

$$g \circ f : A \to C$$

$$g \circ f(x) = g(f(x))$$

Example 1:   Given $f : \mathbf{Z} \to \mathbf{Z}, \quad f(n) = 2n + 1$ and

$$g : \mathbf{Z} \to \mathbf{Z}, \quad g(n) = n^2$$

find the composite function $g \circ f$.

$$g \circ f(n) = g(f(n)) = g(2n + 1) = (2n + 1)^2$$

Note that the calculation proceeds from the inner parentheses out-wards, evaluating f first. The variable n is called a *free* variable.

Example 2:   Let $\mathbf{R}^+$ be the set of positive real numbers.

Given the functions $f, g : \mathbf{R}^+ \to \mathbf{R}^+$ and $f(x) = \lfloor x \rfloor, g(x) = \sqrt{x}$; then $f \circ g \neq g \circ f$.

For $f \circ g$ and $g \circ f$ to be equal it would be necessary to show that $f \circ g(x) = g \circ f(x)$ for all x in $\mathbf{R}^+$. To show that they are not the same function it is sufficient to demonstrate that they are not equal for just one value of x in $\mathbf{R}^+$, that is to give a counter example.

$f \circ g(3.5) = f(g(3.5)) = f(1.8708) = 1$

$g \circ f(3.5) = g(f(3.5)) = g(3) = 1.7321$

Therefore $f \circ g \neq g \circ f$.

Definition 3.6 (Identity function) *The* identity *function I is defined by*

$$I : A \to B$$

$$I(x) = x$$

Theorem 3.1 (Inverse function) *If the function $f : A \to B$ is a one-one function from A onto B then there exists the function f^{-1}*

$$f^{-1} : B \to A$$

called the inverse *of f, for which*

$$f \circ f^{-1} = I = f^{-1} \circ f.$$

Example: The exponential function (e^x) and the logarithmic function $(\ln x)$ are inverses.

For those values of x at which both functions are defined (i.e. for $x > 0$), we have:

$$\ln e^x = x = e^{\ln x}$$

Definition 3.7 (Operation) *An* operation on a set A *is a function*

$$A \times A \times \cdots A \to A$$

with domain $A \times A \times \cdots A$.
  A binary *operation on* A *is a function* $A \times A \to A$.
  A unary *operation on* A *is a function* $A \to A$.

Example 1: Multiplication is a binary operation on the set of real numbers.

Example 2: 'Change sign' is a unary operation on the set of real numbers defined by

$$negation(x) = (-1) \times x \text{ for all } x \text{ in } \mathbf{R}$$

Example 3: Intersection is a binary operation on the power set of a set A.

3.3 RELATIONS

When we observe a link between different objects we may say there is some relationship between them. The common notion of family relationship is easily extended. Consider the 'parent' relation.

Bill(x) is the parent of Sue(y_1), John(y_2), ... is an example where the relation has several pairs $(x, y_1), (x, y_2)$ with the same first element and different second elements. The pairs (Bill, Sue), (Bill, John) belong to the relation 'parent of'. In mathematics the word *relation* may be seen as a generalisation of 'function' in which the restriction to unique second element is removed.

Definition 3.8 (Relation) *A (binary) relation from a set A to a set B is a subset R of $A \times B$. We say x is related to y and write xRy if and only if (x, y) belongs to R. We may say xRy is true if and only if (x, y) belongs to R.*

  A binary relation from A to A is said to be a binary relation on A.

Example 1: $A = \{1, 2, 3, 4\}, B = \{1, 2, 3\}$. The relation R is defined by xRy is true if and only if $x < y$.

$$R = \{(1, 2), (1, 3), (2, 3)\} = \{(x, y) \in A \times B \mid x < y\}$$

Example 2: $A = B = \mathbf{Z}$. R is defined by $xRy = \mathsf{T}$ if and only if $x - y$ is even

$$R = \{(x, y) \in \mathbf{Z} \times \mathbf{Z} \mid x - y = 2k \text{ for some } k\}$$

For this example the set of pairs (x, y) is infinite.

The relation in example 2 is that of 'congruence modulo 2'.

Congruence modulo m

In the exercises at the end of Chapter 1, $\mathbf{Z}_m (m \geq 2)$ was defined to be the set of numbers $\{0, 1, 2, \cdots (m-1)\}$ together with certain binary operations for addition and multiplication. The numbers in $\mathbf{Z}_m$ are, in fact, the possible remainders when an integer is divided by m.

Definition 3.9 *A pair of integers a, b is said to be* congruent modulo m *if and only if they have the same remainder when divided by m.*

We write $a \equiv b \bmod m$ and say a is congruent to b modulo m.

Example 1: $17 \equiv 12 \bmod 5$. (In each case the remainder is 2.)

Example 2: $18 \equiv -3 \bmod 7$. (In each case the remainder is 4.)

Example 3: $16 \equiv 0 \bmod 16; 14 \equiv -2 \bmod 16$. Such properties are used in the twos-complement representation of integers in 4-bit words.

Theorem 3.2 *The integers a, b are congruent modulo m if and only if $a - b = km$ for some integer k.*

Proof: 1. Given that a, b are congruent modulo m, it follows that

$$a = q_1 m + r \quad \text{and} \quad b = q_2 m + r$$

for some integers q_1, q_2, r (definition 3.13)

Therefore $a - b = m(q_1 - q_2)$ where $q_1 - q_2$ is an integer. Choose k to be $q_1 - q_2$, then

$$a - b = km$$

2. Given $a - b = km$.

Divide b by m to obtain $b = qm + r$ and $0 \leq r < |m|$

$$
\begin{aligned}
\text{Then} \quad a - b &= a - (qm + r) \\
&= km \quad \text{(given)} \\
\text{Therefore} \quad a &= (qm + r) + km \\
&= (q + k)m + r
\end{aligned}
$$

Thus a and b have the same remainder (r) when divided by m.

For a given m, if $a \equiv b \bmod m$ then the pairs (a, b) form a subset of $\mathbf{Z} \times \mathbf{Z}$, but the subset is a relation, not a function. A counter example to the proposition that 'congruence modulo 5' is a function could be $17 \equiv 2 \bmod 5$ and $17 \equiv 12 \bmod 5$ which show two congruent pairs with the same first element but different second elements. Congruence modulo m is a binary relation on $\mathbf{Z}$.

Congruence modulo 5 partitions the integers into five classes. The classes are the subsets of the integers congruent to 0,1,2,3,4, respectively. If we carry out operations such as addition or multiplication on these integers modulo 5, then the arithmetic may be carried out using the representatives 0,1,2,3,4 for members of each class. For example:

$$17 + (-2) \equiv 2 + 3 \equiv 5 \equiv 0 \bmod 5$$

3.3.1 Equivalence relations

The relation 'congruence modulo 7' partitions the days of the year into seven classes, namely Sunday, Monday, ... Saturday. For example day 178 and day 241 of the year are the same day of the week since

$$178 \equiv 3 \bmod 7 \text{ and } 241 \equiv 3 \bmod 7$$

(In 1991 day 178 and day 241 were Thursdays.)

On the other hand, the relation 'is a friend of' does not partition a set of people into subsets of friends. For example, suppose a, b are friends; b, c are friends but a, c are not friends. The subsets of mutual friendship groups in $\{a, b, c\}$ are $A_1 = \{a, b\}$ and $A_2 = \{b, c\}$. But this division is not a partition because $A_1 \cap A_2 = \{b\} \neq \emptyset$.

We would like to know what additional properties are needed so that a relation on a set should induce a partition on that set.

To avoid the problem in the 'is a friend of' example we need to have the property 'if a has the relation with b and b has the relation with c then also a has the relation with c'. This property is called the *transitive* property.

In addition, two more properties are required of a relation that induces a partition on its set. Each element of the set must belong to its own cell of the partition, so it must have the relation with itself. This is called the *reflexive* property.

If an element x has the relation with y, then they must be in the same cell of the partition and so y must have the relation with x. This is called the *symmetric* property.

Now to set this down formally:

Definition 3.10 (Equivalence relation) *A binary relation R on a set A is:*

1. reflexive *if xRx for every x in A;*

2. symmetric *if xRy implies yRx for all x, y in A;*

3. transitive *if xRy and yRz implies xRz for all $x, y, z \in A$.*

A relation with all three properties is called an equivalence *relation.*

Definition 3.11 (Equivalence class) *Let R be an equivalence relation on a set A. For each x in A the equivalence class of x under R is defined by*

$$[x] = \{s \in A \mid sRx\}$$

Theorem 3.3 (Partition) *The set of all equivalence classes of A under a given equivalence relation form a partition of A.*

Example 1: The relation 'begin with the same letter' is an equivalence relation on the set of words spelled in English. There are 26 equivalence classes, namely:

'the words beginning with a'

'the words beginning with b'

. . .

and the set of equivalence classes is a partition of the set of words.

Example 2: Congruence modulo p is an equivalence relation.

1. $x \equiv x \bmod p$ since $x - x = 0$ a multiple of p.
   (reflexive property)

2. if $x \equiv y \bmod p$ then $y \equiv x \bmod p$
   (symmetric property)
   for if $x - y = qp$ then $y - x = -qp$ a multiple of p.

3. if $x \equiv y \bmod p$ and $y \equiv z \bmod p$ then $x \equiv z \bmod p$
   (transitive property)

$$
\begin{aligned}
\textbf{Proof:} \quad x - y &= qp \\
y - z &= rp \\
\text{(adding)} \quad x - z &= (q+r)p \quad \text{a multiple of } p \\
\text{That is} \quad x &\equiv z \bmod p
\end{aligned}
$$

Therefore congruence modulo p is an equivalence relation.

The binary relation 'congruence modulo p' on **Z** partitions **Z** into p 'equivalence classes', by theorem 3.3. By theorem 3.2 the equivalence classes are the sets of integers congruent to each of $0, 1, \cdots p - 1$ modulo p.

Example 3: The relation 'equals' is an equivalence relation on the set **Q**.

   The equivalence classes are sets of equal rational numbers.

   For example $\cdots -2/-3, 2/3, 4/6, 6/9, 8/12, \cdots$ is an equivalence class.

Example 4: A relation ρ is defined on the integers by $a\rho b = \mathsf{T}$ if and only if $a - b$ is even. (See example 2, p. 54.) Prove that ρ is transitive.

   Aim: To prove that if $a\rho b$ and $b\rho c$ then $a\rho c$.

   Proof: $a - b$ is even; $b - c$ is even.

> Therefore $(a - b) + (b - c) =$ even $+$ even $=$ even.
> Therefore $a - c$ is even and $a\rho c$.
> Therefore ρ is transitive.

3.3.2   Partial order relations

The idea that objects can be placed in order is an important one in mathematics. The thing to remember is that the primary motivation for the definition of order comes from 'less than or equal to' and not from 'less than'. Paul Halmos (1961, p. 54) says:

> There is no profound reason for this; it just happens that the generalisation of 'less than or equal to' occurs more frequently and is more amenable to algebraic treatment.

Definition 3.12 (Antisymmetric) *A relation R on a set A is* antisymmetric *if for all x, y in A, if xRy and yRx then $x = y$.*

If the relation is reflexive, antisymmetric and transitive then the relation is a *partial order*.

Definition 3.13 (Partial order) *A partial order on a set A is a relation that is reflexive, antisymmetric and transitive. If in addition for every pair x, y in A, xRy or yRx, then R is a* total order *on A.*

Example 1: *Divides* is a partial order on the set $A = \{1, 2, 3, 4, 5, 6\}$. It is not a total order since some pairs are not 'comparable'. For example 3 *divides* 5 is F and 5 *divides* 3 is F.

Example 2: *is a subset of* is a partial order on the power set of $\{a, b, c, d\}$. The relation is not a total order because for a pair of sets such as $\{a, b\}, \{b, c\}$ the relation does not hold in either direction.

Example 3: Prove that *divides* is a total order on the set

$$A = \{1, 2, 4, 8, 16\}.$$

Proof: $b = ma$ and $a = nb$ for some integers m, n.
Therefore $b = mnb$ (substitution) and $mn = 1$.
Now m, n are integers so $m = n = 1$ or $m = n = -1$.
But all members of A are positive, therefore $a = b$.
Therefore *divides* is antisymmetric.
The proof should also establish that *divides* is reflexive and transitive and that for each pair a, b chosen from A, either a divides b or b divides a.

The relation in example 3 is a total order on A.

If a total order is defined on a set then every pair of elements in the set may be compared by deciding which of the two is the larger and which the smaller. Such a set may be called an 'ordered' set, and for a set S with an order relation ρ it would be denoted by (S, ρ). An ordered set of numbers is said to be *continuous* if for every pair of distinct numbers there is a third number which is between them.

Definition 3.14 (Well-ordering principle) *An ordered set is said to be* well-ordered *if and only if every non-empty subset has a least element.*

For example, the ordered set $(\mathbf{N}, \leq)$ is well-ordered. Any subset of the natural numbers has a least element. For example, the set of natural numbers less than 2 has as its least element the number 0.

Now consider the ordered set $(\mathbf{Z}, \leq)$. Some subsets have a least element, for example $\{-2, 3, -4\}$. But there are sets such as, for example, the set of integers less than 2, that do not have a least element. It follows that the set of integers with the order relation 'less than or equal to' is not well-ordered.

The well-ordering property, despite its simplicity, is very important. If an ordered set has this property then it is possible to use mathematical induction, a powerful method of proof for propositions defined on the set.

3.4 SEQUENCES

Two structures that may be used to specify a collection of objects or elements are

- set, and

- sequence.

In a set, *order* is unimportant so $\{a, b, c\} = \{c, a, b\}$. But in a sequence, order is a required property of the collection, and the sequences a_1, a_2, a_3 and a_2, a_3, a_1 are different. We may speak, for example, of a_3 as the third member of the sequence a_1, a_2, a_3, a_4.

A way to convey this idea is to define a sequence as a mapping, or function whose domain is a subset of **N**.

Definition   3.15 (Sequence) *A sequence is a function whose domain is the set of natural numbers,* $\mathbf{N} = \{0, 1, 2, \cdots\}$, *or sometimes* $\{m, m + 1, \cdots\}$ *where m is usually 1.*

A sequence may be finite, *in which case the domain is* $\{m, m + 1, \cdots n\}$ *and again m is usually 0 or 1.*

Some corresponding structures in computing are:

string   a sequence of characters from an alphabet (sometimes called a word).

list   a sequence of records (which may themselves be lists).

record   a sequence of data elements.

text   a sequence of words alternating with characters from { *space, comma space, fullstop space, ...* }.

For some purposes lists and records may be usefully thought of as sets.

Sequences may be described:

1. by writing down the members of the sequence.

   Example: $1, 3, 5, \cdots, 99$ (a finite sequence).

2. by a formula for the general term.

   Example: The sequence of odd numbers $1, 3, 5, \cdots, (2n - 1), \cdots$ or, $a_1, a_2, \cdots, a_n, \cdots$ where $a_n = 2n - 1$ (an infinite sequence).

3. by a recursive procedure.

   Example 1:   The sequence $a_1, a_2, \cdots, a_n, \cdots$ where

$$a_n = \begin{cases} 1 & \text{if } n = 1 \\ a_{n-1} + 2 & \text{otherwise} \end{cases}$$

The formula generates the sequence

$$1, 3, 5, 7, \cdots$$

where each term is obtained from its predecessor in the sequence by adding 2.

Example 2: The sequence

$$x_0, x_1, x_2, \cdots, x_n, \cdots$$

is defined by the formula $x_n = 3x_{n-1} - 2x_{n-2}$ for all $n \geq 1$ and the initial conditions $x_0 = 0$ and $x_1 = 1$. The following table shows the calculation of the first five terms of the sequence.

n	x_n
0	0
1	1
2	$3 \times 1 - 2 \times 0 = 3$
3	$3 \times 3 - 2 \times 1 = 7$
4	$3 \times 7 - 2 \times 3 = 15$
	$\cdots$

The sequence is

$$0, 1, 3, 7, 15, \cdots$$

3.4.1 Strings

A character is a symbol such as any of $a, b, 1, 2, *, (, \cdots$.
An alphabet is a set of characters, for example:

1. the parenthesis alphabet is $\{(,)\}$;

2. the binary alphabet is $\{0, 1\}$;

3. the dictionary alphabet is $\{a, b, c, \cdots z\}$;

4. the character alphabet for a keyboard is the set of characters produced by operating the keys on the keyboard.

A string is a sequence of characters from an alphabet. Normally the characters are placed side by side, not separated by commas (they are said to be *juxtaposed*). The *length* of the string is the number of characters in it. Sometimes a string is called a word, although it need not have any meaning in ordinary language.

Example 1: 110, 00001 and 011010 are strings from the binary alphabet. They have lengths 3, 5 and 6 respectively.

Example 2: *ab, aa, bc* are strings of length 2 from the alphabet $\{a, b, c\}$.

The set of strings of length n from an alphabet A is denoted by A^n.

Example 1: The set of strings of length 2 from the binary alphabet $\{0, 1\}$ is:

$$A^2 = \{11, 01, 10, 00\}$$

Example 2: The set of strings of length 2 from the alphabet $\{a, b, c\}$ is:

$$A^2 = \{aa, ab, ac, ba, bb, bc, ca, cb, cc\}$$

A string of length 0 is defined to be the *empty* string written λ in this book. There is exactly one empty string. Note that λ is not 0 or the 'space' character and $\lambda \notin A$ for any alphabet A. A^0 is the set consisting of λ.

A suitable *universal* set for discussion of strings would be the set of strings of all lengths from a given alphabet.

Definition 3.16 (Closure) *The (Kleene) closure of the alphabet A is the set of strings of all lengths that can be made from the characters of A:*

$$A^* = A^0 \cup A^1 \cup A^2 \cup \cdots \quad = \bigcup_{i=0}^{\infty} A^i$$

The set of all non-empty strings from A is defined to be

$$A^+ = A^1 \cup A^2 \cup \cdots \quad = \bigcup_{i=1}^{\infty} A^i$$

The sets A^*, A^+ are possible universal sets for strings. They are defined constructively so there is no risk of paradox.

A basic operation on strings is to join them together to form new strings.

Definition 3.17 (Concatenate) *The binary operation concatenate on A^* is a function from $A^* \times A^*$ to A^* defined by*

$$\text{concatenate}(\alpha, \beta) = \alpha\beta$$

for all strings $\alpha, \beta \in A^*$.

Example 1: concatenate(011,01) = 01101

Example 2: concatenate(alpha, bet) = alphabet

3.5 EXERCISES

1. (a) Write down the set of characters in the word 'mathematical'.

    (b) Generate all the subsets of $\{a, b\}$.

    (c) Use a predicate to describe the set $\{1, 2, 4, 8, 16, \cdots\}$.

2. Describe the following sets in words. Is it true that $A \cap B = C$? Explain your answer. The universal set is the set of integers **Z**.

$$
\begin{aligned}
A &= \{j \mid j = 8t\} \\
B &= \{j \mid j = 6t\} \\
C &= \{j \mid j = 48t\}
\end{aligned}
$$

3. Find the number of positive integers less than or equal to 100 that are odd or the square of an integer. Is the set of such integers well-defined? Is 25 in the set?

4. Two dice are tossed. Each die is numbered $1, 2, \cdots, 6$ so that the set of possible outcomes is $\{(1, 1), (1, 2), (2, 1) \cdots (6, 6)\}$. What is $n(S)$? Partition S into subsets in which the sum of the two dice is the same.

5. For all sets A, the *power* set of A, written $\mathcal{P}(A)$, is defined to be the set of all subsets of A.

    (a) Given $A = \{a, b, c\}$ find $\mathcal{P}(A)$.

    (b) Given $n(S) = k$ find the number of elements in $\mathcal{P}(S)$.

6. An 80 column, 25 row screen is defined by the set

$$
S = \{(c, r) \mid (c, r) \in \mathbf{N} \times \mathbf{N} \text{ and } 1 \leq c \leq 80 \text{ and } 1 \leq r \leq 25\}
$$

where c represents the column number and r the row number of a 'character position' or 'point' on the screen. S is the universal set for this question.

Windows W_1, W_2 are defined as the sets:

$$
\begin{aligned}
W_1 &= \{(c, r) \mid 20 \leq c \leq 40 \text{ and } 10 \leq r \leq 20\} \\
W_2 &= \{(c, r) \mid 30 \leq c \leq 50 \text{ and } 5 \leq r \leq 15\}
\end{aligned}
$$

Define the sets (a) $W_1 \cap W_2$ and (b) $W_1 - W_2$ in the same way.

7. **(Barber paradox)** In a certain village there is a barber. The barber shaves every man who doesn't shave himself and no one else. Let B be the set of men shaved by the barber. Is B well-defined?

8. Identify the functions among the following relations.

   (a) $\{(0,0),(1,1),(2,2),(3,2)\}$

   (b) $\{(0,1),(0,-1)\}$

9. The function *mean* is defined:

$$mean: \mathbf{N} \times \mathbf{N} \rightarrow \mathbf{N}$$

$$mean(x,y) = \frac{x+y}{2}$$

(a) Evaluate $mean(8,12)$.   (b) Is *mean* a total function?
(c) Find the domain of *mean*.   (d) Is *mean* a one-one function?

10. $\mathbf{N} = \{0,1,2,3,\cdots\}$

$$minus: \;\; \mathbf{N} \times \mathbf{N} \mapsto \mathbf{N}$$

$$minus(x,y) = x - y \text{ if } x \geq y, \text{ 0 otherwise.}$$

Is *minus* a total function? Does it have an inverse? or if not why not? Is the function *into* **N**?

11. (Use a calculator for this question and the next.) For this question define the set $P = \{x \in \mathbf{R} \,|\, x \geq 0\}$. The function f is defined:

$$f : P \rightarrow P$$

$$f(x) = \sqrt{x}$$

The function g is defined:

$$g : \mathbf{R} \rightarrow P$$

$$g(x) = e^x$$

Calculate: (a) $f(7)$ (b) $g \circ f(7)$ (c) $g(7)$ (d) $f \circ g(7)$

12. Define the function f by the calculator function ln, and the function g by the calculator function e^x.

Find $f \circ g(3.6)$ and $g \circ f(3.6)$. Comment on your answer.

13. Explain why, if n is an odd integer, then

$$\lceil (n+1)/2 \rceil = (n+1)/2$$

14. A function $f : A \to B$ is said to be '*one-one*' if and only if,

$$\text{for all } x, y \in A \text{ if } x \neq y \text{ then } f(x) \neq f(y).$$

Show that the function

$$g : \mathbf{Z} \to \mathbf{Z}_8$$

$$g(n) = n \bmod 8$$

is not one-one.

15. The function *successor* is defined

$$successor : \mathbf{N} \to \mathbf{N}$$

$$successor(n) = n + 1$$

Is *successor* a total function? Is *successor* a one-one function? Is the function *onto* **N**? Does *successor* have an inverse?

16. The function g is defined on the set $\mathbf{Z}_7 = \{0, 1, 2, \cdots, 6\}$ by

$$g : \mathbf{Z}_7 \to \mathbf{Z}_7$$

$$g(x) = 2x + 3 \bmod 7$$

(a) Find the value of $g(0), g(1), g(3)$.

(b) Is g a total function?

(c) Is g a one-one function?

(d) Does the function g have an inverse?

(e) Is g onto $\mathbf{Z}_7$?

17. The function f is defined on the set $\mathbf{Z}_{16} = \{0, 1, 2, \cdots, 15\}$ by

$$f : \mathbf{Z}_{16} \to \mathbf{Z}_{16}$$

$$f(x) = x^2 + 5 \bmod 16$$

(a) Find the value of $f(0), f(1), f(4)$.

(b) Is f a total function?

(c) Is f a one-one function?

(d) Does the function f have an inverse?

(e) Is f onto $\mathbf{Z}_{16}$?

18. (a) Show that $51 \equiv 6 \bmod 9$.

   (b) Show that $124 \equiv 7 \bmod 9$.

   (c) Show that $51 + 124 \equiv 6 + 7 \bmod 9$.

   The above exercises illustrate two theorems that you are now asked to prove.

   (d) Prove that, if $u \equiv v \bmod m$ and $t \equiv w \bmod m$
       then $u + t \equiv v + w \bmod m$.

   (e) Prove that, if $u = abc_{10}$ where a, b, c, are the digits of u represented in base 10, then $u \equiv a + b + c \bmod 9$.

   (f) (Optional) Generalise the proposition in (e).

19. (a) Show that $27 \equiv 5 \bmod 11$ and that $35 \equiv 2 \bmod 11$.

   (b) Show that $27 \times 35 \equiv 5 \times 2 \bmod 11$.

   (c) Prove that, if $u \equiv v \bmod m$ and $t \equiv w \bmod m$
       then $u \times t \equiv v \times w \bmod m$.

20. A relation ρ is defined on the integers by $a \rho b = \mathsf{T}$ if and only if $a - b$ is even.

   What properties must ρ have to be an equivalence relation? Prove these properties for ρ.

21. In this question the symbol '$\wedge$' represents 'and', and the symbol '$\vee$' represents 'or'. Let S be the set of screen positions defined by the set

$$S = \{(c, r) \in \mathbf{Z} \times \mathbf{Z} \mid (1 \le c \le 80) \wedge (1 \le r \le 30)\}$$

   Let $\mathbf{B} = \{\mathsf{T}, \mathsf{F}\}$ be the set of Boolean values. A relation *follows* is defined by

$$follows : S \times S \to \mathbf{B}$$

$$(c_1, r_1)\, follows\, (c_2, r_2) = \begin{cases} \mathsf{T} & \text{if } [(c_1 > c_2) \wedge (r_1 = r_2)] \vee (r_1 > r_2) \\ \mathsf{F} & \text{otherwise} \end{cases}$$

   (a) Describe the relation *follows* in words.

   (b) The relation *follows* is not a partial order because it fails two of the three required properties for a partial order. Which of the three properties of a partial order *does* hold for *follows*? Write down a statement of what must be proved to show that *follows* does satisfy the property you claim.

22. The relation 'divides' (written ρ in this question) is defined on the natural numbers by

$$a\rho b = \mathsf{T}$$

if and only if $b = ka$ for some natural number k and $a \neq 0$.

Prove that ρ is antisymmetric.

23. The relation 'power of' (written ρ in this question) is defined on the integers by

$$a\rho b = \mathsf{T}, \text{ if and only if } a = b^n \text{ for some natural number } n.$$

Prove that ρ is antisymmetric. What other properties of ρ would need to be proved to show that it is a partial order?

24. The relation *divides* is defined on the set $\{3, 9, 27\}$; a *divides* b means $b = ma$ for some positive integer m.

Show that *divides* is transitive. Explain why *divides* is a total order.

25. List the first five terms of the sequence $a_1, a_2, \ldots, a_n, \ldots$ where

$$a_n = \begin{cases} 2 & \text{if } n = 0 \\ a_{n-1} + 2 & \text{otherwise} \end{cases}$$

26. List the first five terms of the sequence

$$x_0, x_1, x_2, \cdots, x_n, \cdots$$

defined by the formula $x_n = 3x_{n-1} - 2x_{n-2}$ for all $n \geq 1$ and the initial conditions $x_0 = 0$ and $x_1 = 1$.

27. (a) List A^3, the set of strings of length 3 from the alphabet A, where $A = \{0, 1\}$.

(b) Concatenate the strings $\alpha = \mathtt{cap}$ $\beta = \mathtt{able}$.

28. Sequences of integers $\{s_n\}$ and $\{d_n\}$ are constructed as follows:

$$s_1 = 1; \quad d_1 = 1;$$

$$s_{n+1} = s_n + d_n, \quad d_{n+1} = 2s_n + d_n \text{ for } n = 1, 2, 3\ldots$$

(a) Calculate s_4, d_4

(b) Calculate $r = \dfrac{d_4}{s_4}$

(c) Find the percentage relative error in taking r, from (b) above, as an approximation for $\sqrt{2}$.

29. The universal set for this question is

$$U = \{x \in \mathbf{Z} \mid 1 < x < 101\}.$$

Subsets S_k of U are defined

$$S_k = \{x \mid k \text{ is the smallest prime factor of } x\}$$

(a) Describe the sets S_5, S_{11}.

(b) What is the value of $n(S_{41})$?

(c) Explain why $S_2, S_3, S_5, ... S_{97}$ is a partition of U.

30. Read the following article on sequences.

J.R.Ridenhour, *Mathematics Magazine*, Mathematical Association of America, **59**, 2, 95-105, April 1986.

The following questions should be attempted before commencing Chapter 5.

31. Three consecutive positive integers are multiplied together to give a number n.

Is 2 a factor of n?   Is 3 a factor of n?   What is the largest number that must be a factor of n?

32. Let p be a prime greater than 2. Is $p^2 - 1$ divisible by 2? Is $p^2 - 1$ divisible by 3?

33. What is the largest number that must be a factor of $p^2 - 1$ for all primes p greater than 2?

34. Find a prime number that is one less than a perfect square. Find another prime number that is one less than a perfect square. How many such prime numbers are there?

Chapter 4

Logic

After completing this chapter you should be able to analyse compound propositions, translate between English and symbolic logic expressions, and reason correctly from accepted logical laws. You should be able to use the existential quantifier and the universal quantifier with predicates.

Logic is the science of reasoning. Here we will be primarily concerned with mathematical logic, including an algebra of logical symbols. Modern symbolic logic was set up by George Boole (1815-64) and others to allow the study of reasoning separated from the possible ambiguities and emotive argument that may be present in ordinary language. The idea for a *universal calculus* 'in which all truths of reason would be reduced to a kind of calculus' goes back to Leibniz (1646-1716). This system of symbolic logic and its associated algebra was seized upon by those who constructed the early computers as a model of correct reasoning which could be implemented by electronic circuits.

There are at least six reasons for us to study logic.

- At the hardware level the design of 'logic' circuits to implement instructions is greatly simplified by the use of symbolic logic.

- At the software level a knowledge of symbolic logic is helpful in the design of programs. Many computer languages use Boolean variables and their associated operators. Truth tables are especially useful for deciding if two compound statements are logically equivalent.

- At the level of specification of algorithms symbolic logic contributes as a language of precision. While still in the developmental stage there is some promise that symbolic logic can help with the design, proof of correctness and maintenance of programs.

- Several computer languages are logic-based, for example Prolog. It is believed by many that logic-based languages have advantages for the analysis of information in a database.

- In mathematical reasoning and in the construction of mathematical proofs, a study of logic can be helpful.

- In ordinary reasoning, logic can help to clarify arguments.

4.1   COMPOUND PROPOSITIONS & TRUTH TABLES

The elementary objects of our discussion may be called *statements* or *propositions*. The essential property of a *statement* or *proposition* is that it has attached to it the label (T) or (F), but not both.

Examples: 1. 6 is prime. **2.** 5 is prime. **3.** $(-3 < 2)$. **4.** New York is the capital of the USA. **5.** H_2O is an acid. **6.** The profit on the deal is $10 000.

> The statements in **1.** to **4.** have the labels F, T, T, F. For **5.**, **6.** we would seek advice from a chemist and an accountant respectively.

We are not really interested in the actual truth or falsity of the above statements. We are concerned about how to calculate the truth value of compound statements composed of elementary statements such as those in the six examples. The exact nature of the elementary statements may be left to specialists in the particular field. The statements may be in any 'language', for example English, French, a programming language, a mathematical system or an artificial language. The elementary statements are 'atomic' in the sense that they cannot be broken down to simpler components.

The letters $p, q, r, \cdots$ represent unspecified statements. The symbols $\neg, \vee, \wedge, \rightarrow, \leftrightarrow$ are used to denote operations by which compound statements may be built up from simpler statements. The symbols correspond in a way to be explained to the connectives 'not', 'or', 'and', 'if ... then ... ' and '... if and only if... '.

Compound propositions may be constructed by applying *logical connectives* to one or more given propositions. A logical connective is defined by its truth table. The truth table of a connective shows how the truth value of a compound statement, involving the connective and the elementary propositions so 'compounded' or 'connected', depends on the truth values of those propositions. The compound statement is evaluated for each possible combination of truth values of the atomic statements in the compound. The truth table of a connective is, of course, intended to reflect the natural meaning of the name of the connective. In some cases, however, the same words have different meanings in different contexts, so it is not possible for the truth table definition of a connective to represent every possible meaning associated with its name. The resulting logical algebra is itself perfectly clear, in the sense that it gives a completely reliable system of logic. In the field of computer circuit design, logical algebra has been of great importance.

Example 1: If it rains then the barbecue will be cancelled.

> There is a suggestion that there *will* be a barbecue if it doesn't rain. You might feel that the statement is misleading if it turns out that it doesn't rain and there is no barbecue.

Example 2: If the emergency cord is pulled the train will stop.

> In this example there is no suggestion at all that the train will keep going if the emergency cord is not pulled. The driver may stop the train for a station or a red signal, so the statement gives no indication of what will happen if the emergency cord is not pulled. Therefore we would have no quarrel with the statement if it turned out that the emergency cord was not pulled but the train nevertheless stopped.

The above examples are intended to show that the words 'if ... then ... ' may have slightly different meanings in English, depending on the context. However you will see below that 'if ... then ... ' has just one precise meaning as defined in our logic. It is this meaning that is used in a computer to evaluate compound statements involving logical connectives.

While as a matter of honesty it is important to point out that the logical connectives we are about to define do not equate exactly to every usage of the corresponding words in English, experience shows that the connectives and the associated logical algebra is quite useful in many mathematical and computing contexts.

not (negation)

We will use $\neg p$ to represent the negation of the statement p. 'Negation' is defined by the truth table below.

p	$\neg p$
T	F
F	T

Negation in our algebra is intended to be a model for negation or denial in ordinary language. Thus the symbol $\neg$ may conveniently be read as 'not', T may be read as 'true' and F may be read as 'false'. All possible combinations of truth values are listed in the column on the left under the p. The column on the right shows the corresponding values of the proposition $\neg p$. There are two lines for the table since the elementary (atomic) statement p has two possible truth values.

For example, if p represents the statement $(-3 < 2)$ then $\neg p$ represents the statement $(-3 \geq 2)$. In this case p is true while $\neg p$ is false, and this case is covered by the first line of the truth table.

Example: Show that p and $\neg(\neg p)$ have the same truth table.

p	$\neg p$	$\neg(\neg p)$
T	F	T
F	T	F

The statements p and $\neg(\neg p)$ have identical truth values on each line of the truth table (compare columns 1 and 3). We say that $\neg(\neg p)$ is *logically equivalent* to p.

The statement $\neg(\neg p)$ is normally written $\neg\neg p$, since there should be no confusion about what is meant.

Negation may be regarded as a function that maps *statements* to *statements*.

and (conjunction)

We will use $p \wedge q$ to represent the conjunction(i.e. 'and') of the statements p, q. Conjunction is defined by the truth table below.

p	q	$p \wedge q$
T	T	T
T	F	F
F	T	F
F	F	F

Conjunction in our algebra is intended to be a model for the connective 'and' in ordinary language. Thus the symbol $\wedge$ may conveniently be read as 'and'.

There are four possible truth combinations and the truth table must have four lines. The operator $\wedge$ maps pairs of statements to statements, thus it is a binary operator on statements.

Observe that by exchanging p, q the truth values for '$\wedge$' are not changed. The statements $p \wedge q$ and $q \wedge p$ are logically equivalent.

In ordinary language there may be a difference in meaning between usages of 'and'.

Compare

'she developed a fever and drank a glass of water'

with

'she drank a glass of water and developed a fever'.

The change in order appears to change the meaning of the statement because we associate a time with each event in the order in which it is mentioned. We at least suspect that the first mentioned event influenced the occurrence of the second. Therefore this particular usage of 'and' is not well modelled by our connective $\wedge$.

Example 1:   Show that $p \wedge p$ is logically equivalent to p.

p	$p \wedge p$	p
T	T	T
F	F	F

line 1: When p is T $p \wedge p$ is T $\wedge$ T which is T from the truth table for 'and'. The value T is placed under the '$\wedge$' in $p \wedge p$.

line 2: When p is F, $p \wedge p$ is F $\wedge$ F which is F.

The truth table for $p \wedge p$ is the same as for p, so $p \wedge p$ is logically equivalent to p.

Example 2:   Show that $p \wedge \neg p$ is logically false. By this we mean that the truth value of the statement $p \wedge (\neg p)$ is F on each line of the truth table.

p	(p)	$\neg p$	$(p \wedge \neg p)$
T	T	F	F
F	F	T	F

The parentheses about $\neg p$ have been dropped since it is clear that $\neg$ refers to the statement immediately following. In the truth table we evaluate the statements in order left to right; first p, then $\neg p$, then $p \wedge \neg p$. For each value of p, $p \wedge \neg p$ is F. So we conclude that $p \wedge \neg p$ is logically false.

A compound statement that has the value F for all possible values of the elementary statements in it is called a *contradiction*.

The next two operators model different usages of the word 'or'. In the context 'one or the other or both', we say the *inclusive or* is being used. In the context 'either one or the other but not both', we say the *exclusive or* is being used.

or (disjunction)

We will use $p \vee q$ to represent the 'inclusive or' of the statements p, q. The 'inclusive or' operator is defined by the truth table below.

p	q	$p \vee q$
T	T	T
T	F	T
F	T	T
F	F	F

The 'inclusive or' in our algebra is intended to be a model for the connective 'or' in ordinary language. The symbol $\vee$ may conveniently be read as 'or'. The connective $\vee$ defined above models a common usage of the word 'or', illustrated in the following sentence.

'If the control file is missing or the data file is missing then send the message "file(s) missing" and abort the session.'

We would like the indicated action to take place if either file is missing and also if both files are missing.

The operator $\vee$ maps statement pairs to statements.

Example 1:   Show that $p \vee \neg p$ is logically true. By this we mean that the truth value of the statement $p \vee \neg p$ is T on each line of the truth table. (In the next section this result will be referred to as the law of the excluded middle.)

p	(p)	$\neg p$	$(p \vee \neg p)$
T	T	F	T
F	F	T	T

Evaluate p, then $\neg p$, then $p \vee \neg p$. For each value of p, $p \vee \neg p$ is T. We conclude that $p \vee \neg p$ is logically true.

A compound statement that has value T for all possible values of the elementary statements in it is called a *tautology*.

Example 2:   De Morgan's Laws. Show that:

   (a) $\neg(p \wedge q)$ is logically equivalent to $\neg p \vee \neg q$.
   (b) $\neg(p \vee q)$ is logically equivalent to $\neg p \wedge \neg q$.

(a)

			(left)				(right)
p	q	$p \wedge q$	$\neg(p \wedge q)$	$\neg p$	$\neg q$	$(\neg p) \vee (\neg q)$
T	T	T	F	F	F	F
T	F	F	T	F	T	T
F	T	F	T	T	F	T
F	F	F	T	T	T	T

$\neg(p \wedge q)$ is logically equivalent to $\neg p \vee \neg q$, each having truth table FTTT.

(b)

			(left)				(right)
p	q	$p \vee q$	$\neg(p \vee q)$	$\neg p$	$\neg q$	$(\neg p) \wedge (\neg q)$
T	T	T	F	F	F	F
T	F	T	F	F	T	F
F	T	T	F	T	F	F
F	F	F	T	T	T	T

$\neg(p \vee q)$ is logically equivalent to $\neg p \wedge \neg q$, each having truth table FFFT

exclusive or

We will use $\underline{\vee}$ to represent the 'exclusive or' connective. It is defined by the truth table below.

p	q	$p \underline{\vee} q$
T	T	F
T	F	T
F	T	T
F	F	F

The 'exclusive or' connective is intended to be a model for the usage of the word 'or' demonstrated in the following sentence.

'Either you are coming or you aren't, make up your mind.'

The exasperated speaker demands that a decision be made between alternatives.

The following mathematical example emphasises that the 'exclusive or' is intended by adding the phrase 'but not both'.

'If either $x < 0$ or $y < 0$, but not both, then $xy < 0$.'

if p then q (conditional)

We will use $p \rightarrow q$ to represent the conditional 'if p then q' or 'p only if q' for the statements p, q. The conditional operator is defined by the truth table below.

p	q	$p \to q$
T	T	T
T	F	F
F	T	T
F	F	T

The conditional operator in our algebra is intended to be a model for the conditional of ordinary language. However some readers may query line 3. Consider the following argument.

Prove 'if $(0 = 1)$ then $(1 = 1)$'.

Proof:

$$
\begin{aligned}
\text{We are given} \quad 0 &= 1. \\
\text{Therefore} \quad 1 &= 0, \\
\text{and by addition} \quad 1 &= 1.
\end{aligned}
$$

The statement $(0 = 1)$ is F and the statement $(1 = 1)$ is T, and the proof above shows that 'if F then T' is T. The argument agrees with line 3 of the truth table for the conditional.

The conditional is used extensively in argument so we should look at some associated forms.

p	q	conditional $p \to q$	converse $q \to p$	contrapositive $\neg q \to \neg p$	inverse $\neg p \to \neg q$
T	T	T	T	T	T
T	F	F	T	F	T
F	T	T	F	T	F
F	F	T	T	T	T

We observe that the converse is *not* logically equivalent to the conditional.

And the contrapositive *is* logically equivalent to the conditional.

If a conditional is true then we can say nothing about the converse in general. The converse may be true or false depending on the particular proposition.

Example (to illustrate the forms associated with the conditional):

Let p be the statement 'I live in Canberra'.

Let q be the statement 'I live in Australia'.

The conditional $p \to q$ says, 'If I live in Canberra then I live in Australia' and we agree this is true (Canberra is the capital of Australia).

The converse $q \to p$ says, 'If I live in Australia then I live in Canberra' and that does not follow at all. Many people live outside Canberra yet still in Australia.

However the contrapositive $\neg q \to \neg p$ says 'If I don't live in Australia then I don't live in Canberra', which is correct and logically equivalent to the conditional above, which places Canberra inside Australia.

Another logical equivalent to the conditional $p \to q$ is the statement $\neg p \vee q$. This equivalence is very useful, although not intuitively obvious. The truth table below establishes it.

p	q	$p \to q$	$\neg p \vee q$
T	T	T	T
T	F	F	F
F	T	T	T
F	F	T	T

p if and only if q (biconditional)

The symbols '$p \leftrightarrow q$', below, are read 'p if and only if q'.

The biconditional is logically equivalent to the compound statement $(p \to q) \wedge (q \to p)$. That is the biconditional is logically equivalent to the conditional and its converse. The truth table for the biconditional follows.

p	q	$p \leftrightarrow q$
T	T	T
T	F	F
F	T	F
F	F	T

Alternative ways of stating the conditional

Perhaps because we are fond of argument there are many phrases that convey the meaning of the conditional. The conditional $p \to q$ may be given as:

if p then q, p only if q, q if p, all p are q, p is sufficient for q or q is necessary for p.

The biconditional $p \leftrightarrow q$ may be read as 'p if and only if q' or 'p is necessary and sufficient for q'.

Omission of brackets (precedence of connectives)

In ordinary algebra we rank the symbols

$$+, -,$$

$$*, \div$$

in order to avoid excessive use of brackets. Thus $-3 + 4 * 5$ means $(-3) + (4 * 5)$, i.e. 17 and not $(-3 + 4) * 5$, i.e. 5

We will do the same here informally, without setting out mechanical rules. The ranking is

$$\neg,$$

$$\lor, \land,$$

$$\rightarrow, \leftrightarrow,$$

from lowest to highest. The rule is that the lowest ranking connectives extend over the shortest range. To fill in missing brackets, start with the lowest ranking connective and insert the nearest brackets permissible, without breaking up existing brackets, then move to the next lowest ranking connective, and so on.

Examples:

1. $\neg p \lor q$ means $((\neg p) \lor q)$.

2. $\neg p \lor q \rightarrow r \land s$ means $(((\neg p) \lor q) \rightarrow (r \land s))$.

3. $\neg p \land q \lor r$ is ambiguous. Brackets are required to show what is meant.

4. $p \land q \land r$ is not allowed (until section 4.2.1).

5. $p \rightarrow q \rightarrow r$ is not allowed.

Note: In computing courses '$\land$' is identified with '$\times$' and '$\lor$' is identified with '$+$' and it is common to give $\times$ precedence over $+$ by analogy with ordinary arithmetic. Thus computing students might not feel that item 3 is ambiguous.

4.2 REASONING FROM LOGICAL ASSUMPTIONS

A *calculus* of propositions is a process by which we reason or compute using symbols. For a 'calculus' we need 'axioms' or assumed facts and 'rules' for the manipulation of axioms to produce new facts. An example of an axiom is '$p \lor (\neg p)$ is logically true for all propositions p'. An example of a rule is the 'rule of replacement' which allows us to replace any proposition by a logically equivalent proposition to obtain a new proposition (or 'theorem'). Using the rule of replacement, replace p by $\neg p$ in the axiom above. Then $\neg p \lor p$ is logically true, a new proposition.

This method of reasoning is more dynamic than the truth table method introduced in section 4.1, but is consistent with that approach in the sense that every axiom that we assume could be established via a truth table.

Definition 4.1 (Tautology) *A compound proposition that has value* T *for all values of the propositions in it is called a* tautology.

Definition 4.2 (Contradiction) *A compound proposition that has value* F *for all values of the propositions in it is called a* contradiction.

Definition 4.3 (Contingency) *A compound proposition that has value* T *for some values of the propositions in it and the value* F *for the rest is called a contingency.*

Definition 4.4 (Logically equivalent) *Propositions* p, q *are said to be* logically equivalent *if and only if* $p \leftrightarrow q$ *is a tautology.*

The definitions above arise from the truth table approach adopted in section 4.1. However we might reasonably accept that propositions such as

'a proposition is logically equivalent to itself' $\hspace{2cm}$ (1)

and

'either a proposition or its negation is true' $\hspace{2cm}$ (2)

are universally true, without feeling any need to explore all the truth table possibilities.

Adopting this approach, 'logically equivalent to' becomes a fundamental relation defined on the set of all propositions that may be constructed from elementary propositions $p, q, r, \ldots$, using the logical connectives $\neg, \wedge, \vee, \rightarrow, \leftrightarrow$, introduced in section 4.1.

A *tautology* is now a proposition which is 'logically equivalent to' T.

A *contradiction* is a proposition which is 'logically equivalent to' F.

Propositions $(1), (2)$ above become:

p $\quad$ 'is logically equivalent to' $\quad$ p $\hspace{3cm}$ (1′)

$p \vee \neg p$ $\quad$ 'is logically equivalent to' T $\hspace{3cm}$ (2′)

Now it would soon become tedious to repeat the phrase 'is logically equivalent to' throughout our reasoning, so it is replaced by the symbol '$\equiv$'. Some writers use the symbol '$\Leftrightarrow$'. The symbol '$\equiv$' here represents a relation between statements. (Note that in the sentence '$a \equiv b \bmod m$' where a, b, m are integers, the symbol represents 'congruence modulo m'. The symbol $\equiv$

is not defined in isolation, but only with respect to a given set of objects, for example statements or integers.)

The propositions 1,2 above are now written:

     Law of identity $p \equiv p$ (1″)

     Law of the excluded middle $p \vee \neg p \equiv \mathsf{T}$ (2″)

It is worth noting that the relation 'is logically equivalent to', written '$\equiv$' is an *equivalence* relation on the set of all propositions that may be constructed from elementary propositions. Equivalence relations were mentioned in section 3.2.

4.2.1 Laws of logical algebra or propositional calculus

The laws given below may be assumed correct and may be used in reasoning. The laws are consistent with the truth table definitions of connectives given in section 4.1.

When applying these laws, one may substitute arbitrary propositions for the letters p, q, r. A logician would say each law is a *schema* covering infinitely many instances.

Laws of negation

     (a) $\neg\neg p \equiv p$ (b) $\neg\mathsf{T} \equiv \mathsf{F}$ (c) $\neg\mathsf{F} \equiv \mathsf{T}$

Commutative laws

     (a) $p \vee q \equiv q \vee p$ (b) $p \wedge q \equiv q \wedge p$

Associative laws

     (a) $p \vee (q \vee r) \equiv (p \vee q) \vee r$ (b) $p \wedge (q \wedge r) \equiv (p \wedge q) \wedge r$

Distributive laws

     (a) $p \wedge (q \vee r) \equiv (p \wedge q) \vee (p \wedge r)$ (b) $p \vee (q \wedge r) \equiv (p \vee q) \wedge (p \vee r)$

De Morgan's laws

     (a) $\neg(p \vee q) \equiv \neg p \wedge \neg q$ (b) $\neg(p \wedge q) \equiv \neg p \vee \neg q$

Law of the excluded middle $p \vee \neg p \equiv \mathsf{T}$

Law of contradiction $p \wedge \neg p \equiv \mathsf{F}$

Law of implication $(p \rightarrow q) \equiv (\neg p \vee q)$

Law of equivalence $(p \leftrightarrow q) \equiv (p \rightarrow q) \wedge (q \rightarrow p)$

Laws of or-simplification

     (a) $p \vee p \equiv p$ (b) $p \vee \mathsf{T} \equiv \mathsf{T}$ (c) $p \vee \mathsf{F} \equiv p$ (d) $p \vee (p \wedge q) \equiv p$

Laws of and-simplification

     (a) $p \wedge p \equiv p$ (b) $p \wedge \mathsf{T} \equiv p$ (c) $p \wedge \mathsf{F} \equiv \mathsf{F}$ (d) $p \wedge (p \vee q) \equiv p$

Law of identity $p \equiv p$

Some of the above laws may be proved from others in the set. It is possible to find *minimum* sets of logical laws from which all others may be derived, but this is not important to us. The laws above have been selected as a reasonable collection on which to base our arguments. More laws could have been given, but at the expense of making the list so long that it would be tedious for reference.

4.2.2 Rules for replacement and transitivity

The rules show how new propositions may be *inferred*.

Rule of replacement This rule allows us to replace any proposition by a logically equivalent proposition to obtain a new theorem.

> **Example 1:** Prove that $(x \to y) \equiv (\neg y \to \neg x)$ (the *conditional* is logically equivalent to the *contrapositive*).
>
> $$\begin{aligned} \textbf{Proof: } (x \to y) &\equiv \neg x \lor y & \text{(implication)} \\ &\equiv y \lor \neg x & \text{(commutative axiom)} \\ &\equiv \neg\neg y \lor \neg x & \text{(negation)} \\ &\equiv (\neg y \to \neg x) & \text{(implication)} \end{aligned}$$
>
> **Example 2:** Show
>
> $$((p \to q) \land p) \to q$$
>
> is a tautology.
>
> $$\begin{aligned} \textbf{Proof: } \quad ((p \to q) \land p) \to q & \\ &\equiv \neg((p \to q) \land p) \lor q & \text{(implication)} \\ &\equiv (\neg(p \to q) \lor (\neg p) \lor q & \text{(de Morgan)} \\ &\equiv \neg(p \to q) \lor (\neg p \lor q) & \text{(associative)} \\ &\equiv \neg(p \to q) \lor (p \to q) & \text{(implication)} \\ &\equiv \mathsf{T} & \text{(excluded middle)} \end{aligned}$$
>
> which proves that $((p \to q) \land p) \to q$ is a tautology.
>
> **Example 3:** Show that $p \land q \to p$ is a tautology.
>
> $$\begin{aligned} \textbf{Proof: } (p \land q \to p) &\equiv \neg(p \land q) \lor p & \text{(implication)} \\ &\equiv (\neg p \lor \neg q) \lor p & \text{(De Morgan)} \\ &\equiv p \lor (\neg p \lor \neg q) & \text{(commutative)} \\ &\equiv (p \lor \neg p) \lor \neg q & \text{(associative)} \\ &\equiv \mathsf{T} \lor \neg q & \text{(excluded middle)} \\ &\equiv \mathsf{T} & \text{(or-simplification)} \end{aligned}$$
>
> Thus $p \land q \to p$ is a tautology.

Example 4: Show that $p \rightarrow (p \vee q)$ is a tautology.

$$
\begin{aligned}
\textbf{Proof:} \quad (p \rightarrow (p \vee q)) &\equiv \neg p \vee (p \vee q) && \text{(implication)} \\
&\equiv (\neg p \vee p) \vee q && \text{(asociative)} \\
&\equiv \mathsf{T} \vee q && \text{(excluded middle)} \\
&\equiv \mathsf{T} && \text{(or-simplification)}
\end{aligned}
$$

Thus $p \rightarrow (p \vee q)$ is a tautology.

Examples 3 and 4 suggest another relation betweeen compound statements or propositions. Where $p \rightarrow q$ is a tautology, we may say p 'logically implies' q. Some writers use the symbol $\Rightarrow$ to mean 'logically implies'.

The relation 'logically implies' is transitive, but not symmetric.

Rule of transitivity If $p \Rightarrow q$ and $q \Rightarrow r$ then $p \Rightarrow r$.

Example: Prove that $p \wedge q \Rightarrow p \vee q$.

$$
\begin{aligned}
\textbf{Proof:} \quad p \wedge q &\Rightarrow p && \text{(example 3)} \\
p &\Rightarrow p \vee q && \text{(example 4)} \\
p \wedge q &\Rightarrow p \vee q && \text{(rule of transitivity)}
\end{aligned}
$$

The logical equivalence

$$\text{proposition}(1) \equiv \text{proposition}(2)$$

may be regarded as the same as

$$(\text{proposition}(1) \Rightarrow \text{proposition}(2)) \wedge (\text{proposition}(2) \Rightarrow \text{proposition}(1))$$

4.3 PREDICATES & QUANTIFIERS

The propositional calculus of logic developed so far in this chapter is a self-contained theory of logic, but it has limitations.

It was declared at the beginning that a statement or proposition p must be unambiguously true or false. But frequently we wish to consider 'statements' whose truth value changes. For example when explaining the conditional we chose p to be 'I live in Canberra'.

For different persons regarding themselves as 'I', p will be true or false. The truth value of p is *predicated* on the 'I' in the sentence 'I live in Canberra'.

We can handle this situation by introducing functions that take truth values as values. Such a function is called a *predicate*. In the above example, the domain variable 'I' has values chosen from a set of people. Each person maps to T if they live in Canberra or F if they do not.

Predicate calculus

Definition 4.5 (Predicate) *A predicate is an expression involving one or more variables defined on some domain. Substitution of a particular value for the variable(s) from the domain produces a proposition which is true or false. Thus a predicate defines a function from a domain to the Boolean set { T, F }.*

Example 1: $P(n) \equiv (n$ is prime$)$ is a predicate on the natural numbers. We observe that $P(1)$ is false, $P(2)$ is true, $P(12)$ is false, $P(37)$ is true.

The predicate 'n is prime' may be considered as a function:

$$P : \mathbf{N} \to \mathbf{B}$$

$$P(n) = \begin{cases} \mathsf{T} & \text{when } n \text{ is prime} \\ \mathsf{F} & \text{when } n \text{ is 1 or composite} \end{cases}$$

Example 2: Preconditions and postconditions are predicates. For example the postcondition for integer division (section 1.1) is $Q(a, b, q, r) \equiv (a = bq + r) \wedge (0 \le r < b)$. This predicate defines the function

$$postdivision : \mathbf{N} \times \mathbf{N} \times \mathbf{N} \times \mathbf{N} \to \mathbf{B}$$

$$postdivision(a, b, q, r) = \begin{cases} \mathsf{T} & \text{when } (a = bq + r) \wedge (0 \le r < b) \\ \mathsf{F} & \text{otherwise} \end{cases}$$

Quantifiers

Apart from substitution of a particular value for the variable there is another important way to convert a predicate to a proposition. That way is by means of a quantifier, which may be used to specify whether the predicate is true on all or part of the domain.

'For all' n, $P(n)$ means that the predicate P has value T across the whole domain. The symbol that we will use to represent 'for all' is '$\forall$'.

'For some' or 'there exists' n $P(n)$ means that the predicate has value T for at least one of the elements of the domain. The symbol that we will use to represent 'for some' is '$\exists$'.

The following are examples of the use of quantifiers, with the truth value that results.

1. The phrase '$2k$ is even' looks at first like a predicate, but is really a statement (T) because it is understood that $2k$ is even for all values of the integer k. In symbols we write

$$\forall k \in \mathbf{N} \; (2k \text{ is even})$$

and the proposition has value T.

2. For all integers n, $2n - 1$ is odd. In symbols we write

$$\forall n \in \mathbf{Z} \; (2n - 1 \text{ is odd})$$

and the proposition has value T.

3. For all primes p, p is odd. In symbols we write

$$\forall p \in \text{the set of prime numbers}, \; (p \text{ is odd})$$

This proposition has value F. (Remember that 2 is prime.)

The quantifier may be restricted to a subset of the domain of the predicate.

4. For every non-zero real number, the square of the number is positive. In symbols,

$$\forall x \in \mathbf{R}, x \neq 0 \; (x^2 > 0)$$

The proposition has value T.

5. If an integer is even then its square is even.

This is really a general statement about all integers. It would be more satisfactory to include the quantifier explicitly. Thus:

For all integers n, if n is even then n^2 is even.

In this form it is clear just one false value of the predicate for a particular value of n would be sufficient to disprove the proposition. However such a *counter example* cannot be found as the following proof shows.

Proof: For all integers n,
$$n \text{ is even implies that } (n = 2m) \text{ for some integer } m$$
$$\text{which implies } n^2 = 4m^2, \quad \text{thus } n^2 \text{ is even.}$$

6. For all integers n, if n^2 is odd then n is odd.

This proposition is the contrapositive of example 5.

The structure of the statements in **5.** and **6.** above may be more transparent if we look at them in symbolic form.

Define P(n) to be the predicate 'n is even'
  Q(n) to be the predicate 'n^2 is even'.
Then the proposition in **5.** above is

$$\forall n \in \mathbf{Z}\,(P(n) \rightarrow Q(n));$$

and the contrapositive in **6.** is

$$\forall n \in \mathbf{Z}\,(\neg Q(n) \rightarrow \neg P(n))$$

Negation of quantified predicates

Since a quantified predicate is a proposition with a definite truth value it must be possible to negate it to obtain the opposite truth value. The symbolic structure of negation is investigated using an example.

For some real number x, x^2 is negative. In symbols $\exists x \in \mathbf{R}\,(x^2 < 0)$, which is F.

The negation of the proposition is:

for all real numbers x, x^2 is not negative. In symbols $\forall x \in \mathbf{R}\,(x^2 \geq 0)$, which is T.

The general results are:

$$\neg(\forall x\ P(x)) \equiv \exists x\,(\neg P(x))$$

$$\neg(\exists x\ Q(x)) \equiv \forall x\,(\neg Q(x))$$

Quantifiers for predicates with more than one variable

The results for negation extend to quantifiers for predicates with more than one variable. For example:

$$\forall x \in \mathbf{R}\ \exists y \in \mathbf{R}\ (x + y = 0)$$

by which we mean ' for every real x there is a number y such that $x + y = 0$'. This is true, the required number y being the negative of x. The negation is:

$$\neg(\forall x \exists y\,(x + y = 0)) \equiv \exists x \neg(\exists y\,(x + y = 0)) \equiv \exists x \forall y\,(x + y \neq 0)$$

The **order** of multiple quantifiers for predicates with more than one variable may change the meaning and the truth value of the proposition. For example:

Every natural number has a successor. In symbols

$$\forall n \in \mathbf{N} \, \exists m \in \mathbf{N} \, (m = n + 1)$$

and this proposition is true.

If we change the order of the quantifiers we get

$$\exists m \in \mathbf{N} \, \forall n \in \mathbf{N} \, (m = n + 1)$$

In ordinary language this says 'some natural number has every natural number as its successor', which is false and not equivalent to the proposition with the quantifiers in the original order.

4.4 EXERCISES

1. The *nand* operator, denoted by '|' is defined by

$$p \mid q \equiv \neg (p \wedge q)$$

(a) Give the truth table for $p \mid q$.

(b) Show that $p \mid p$ is logically equivalent to $\neg p$.

(c) Find a formula for $p \wedge q$ using only the nand operator.

(d) Find a formula for $p \vee q$ using only the nand operator.

2. Show that $p \to q$ is logically equivalent to $\neg p \vee q$, using truth tables.

3. Show that $(x \wedge y) \to x$ is a tautology using truth tables.

4. These four statements carry the same meaning in mathematical usage:

if p then q p is sufficient for q

q is necessary for p p only if q.

Rewrite each of the following statements in three equivalent ways using the above patterns:

(a) If 9 divides n then 3 divides n.

(b) $x = 0$ is sufficient for $xy = 0$.

5. Find the *converse* and the *contrapositive* of the following statements.

(a) If n is odd then n^2 is odd.

(b) If n is prime and n is greater than 2 then n is odd.

(c) If $ab = 0$ then $a = 0$ or $b = 0$.

6. Show that $(p \lor q) \land (\neg p \lor q) \leftrightarrow q$ is a tautology.

7. Use the laws of propositional calculus (section 4.2.1), other than $p \lor (p \land q) \equiv p$ and the proposition itself, to prove $p \land (p \lor q) \equiv p$.

8. Use the laws of propositional calculus (section 4.2.1) to prove that the following are tautologies.

(a) $(p \land q) \rightarrow (p \lor q)$

(b) $(p \rightarrow (q \land r)) \rightarrow ((p \rightarrow q) \land (p \rightarrow r))$

(c) $(p \rightarrow (p \land q)) \rightarrow (p \rightarrow q)$

(d) $(\neg(p \land q) \land p) \rightarrow \neg q$

(e) $((p \lor q) \land \neg p) \rightarrow q$

9. Prove, using the table of logically equivalent propositions, that the following statements are tautologies.

(a) $p \rightarrow ((q \lor r) \rightarrow p)$

(b) $(p \rightarrow (q \rightarrow r)) \leftrightarrow (q \rightarrow (p \rightarrow r))$

(c) $(p \land (p \rightarrow q)) \rightarrow p$

(d) $((p \land q) \leftrightarrow p) \rightarrow (p \rightarrow q)$

(e) $((p \lor q) \land q) \leftrightarrow q$

10. $P(n)$ is the predicate '*if 4 divides n then 2 divides n*'.

What is the truth value of

(a) $P(12)$?   (b) $P(10)$?   (c) $\exists_n P(n)$?   (d) $\forall_n P(n)$?

11. What is the truth value of this statement?

$$\forall_x \exists_y (y - x = x - y)$$

Give the negation of the statement.

12. Write using mathematical and logical symbols the following statement:

For every positive real number x, there is a natural number n such that x equals 2^n or x lies between 2^n and the next largest power of 2.

Write down the negation of the statement in symbols. Which statement is true?

13. In this question the universal set is $\mathbf{Z}$; that is, n is an integer.

$$P(n) \quad \text{is the predicate} \quad (0 < n^2 \le 4)$$
$$R(n) \quad \text{is the predicate} \quad (0 < n^3 \le 8)$$
$$S(n) \quad \text{is the predicate} \quad (0 < n \le 2)$$

Give the set of values of n which generate true statements for each of P, R, S.

Which of the predicates are equivalent?

Is the statement $\forall_n (R(n) \to P(n))$ true?

14. Use the calculus of propositions to simplify each of the propositions below to one of the propositions $F, T, p, q, p \wedge q, p \vee q$.

(a) $p \vee (q \vee p) \vee \neg q$ (b) $p \vee q \vee \neg p$ (c) $\neg p \to (p \wedge q)$

15. *Conjunctive normal form* has the form

$$p_0 \wedge \cdots \wedge p_n \text{ where each } p_i \text{ has form } e_0 \vee \cdots \vee e_m.$$

Each e_j is T, F, or an elementary proposition x or its negation $\neg x$.

Convert the following to conjunctive normal form:

(a) $(p \wedge q) \vee (p \wedge \neg r)$ (b) $(p \to q) \leftrightarrow (\neg p \vee q)$

Chapter 5

Proof

This chapter is about the construction of mathematical proof. You will be expected to complete proofs in which some assistance is provided for the structure of the proof. At a higher level it is important to learn how to select the structure for the proof. The proof structures that you should be able to read and use are: proof by cases, direct proof following analysis, proof by contrapositive, proof by contradiction and mathematical induction. A specialised form of proof has been developed to establish that a program is correct. This method is applied to proving algorithms correct in Chapter 11.

Proof has been an important part of mathematics since the time of the ancient Greeks. Reasoning was undoubtedly part of ancient Egyptian and Babylonian culture but there is no evidence of 'proof' meaning 'a demonstration of an argument in a public way, sufficiently rigorous to withstand criticism'. In a book called *The Elements*, Euclid (*c*.300 BC) presented the concept that is still the basis of proof in mathematics. He demonstrated that if we assume a small number of axioms and accept definitions and rules of argument, then it is possible to prove a rich collection of propositions as consequences of the assumptions. However 'proof' is a feature of many activities. In the courts lawyers present a public argument and in all professions 'proof' plays a part.

In recent years some mathematical methods of proof have been adapted to the task of proving programs correct. Patiently testing a program for many different sets of input data cannot normally guarantee that a program is correct for all possible inputs, since a useful program will be designed to accept an enormous number of possible input sets. It is more satisfying to *prove* the program correct, although this process is likely to be longer and more difficult than constructing the program itself. Proving a program correct is called *verifying* the program.

The efforts at proving programs correct have had an effect on the style of program writing. Some authors advocate the development of a program and its proof side by side.

5.1 PATTERNS OF PROOF

In this section we present a systematic approach to mathematical proof. The proof of propositions is often difficult for students. It requires a very good knowledge of theorems, definitions and assumptions in the area of the proposition, and the ability to select a proof strategy and follow it logically, coping with a variety of difficulties on the way. Yet it is very satisfying to complete a proof and to convince your audience that the proposition is correct beyond doubt.

The motivation for the approach used in this book is that the art of constructing a proof is similar to that of creating a good program in two respects. Each must be correct; and each needs to be understood by others.

The following examples are chosen to illustrate well-known methods of proof. The propositions have been chosen to be about a familiar part of mathematics, the natural numbers. The objective is to keep your attention on the method.

It should be made clear at the beginning that the 'methods' spoken of are not recipes or procedures that will inevitably generate a correct proof for every proposition. The methods are approaches that are likely to succeed, but they require many small variations to suit particular propositions. To cope with this you need to try as many exercises as possible to gain experience.

Those of us who enjoy proving propositions have learned that a good idea may not succeed and you have to be prepared to try another if the first idea doesn't work.

When writing a proof of a proposition we should keep in mind that we are trying to convince some audience that the proposition is true. So our proof must be correct *and* understandable. The ideal would be to draft a proof then try it out on someone else, finally polishing the proof for clarity. Because a proof is meant to be read you may find a good proof spoken of as elegant or lucid. On the other hand, a proof may be difficult to read due to cumbersome notation, or difficult structure. This part of mathematics probably benefits more than any other from interaction between you and a tutor and other students.

The phases we pass through in constructing a proof are:

1. assembling the facts likely to be useful in the proof;

2. choosing a proof strategy or method; and

3. preparing the proof for an audience.

In this section we will put most effort into the second phase. One of the features of proofs as presented in texts is that the reason for taking a particular approach is normally omitted. Often we read a proof and say to ourselves, 'I can see that is true, but how did the writer think of it? '

To help with this problem, comments [in brackets] will be placed in the proofs in this section to suggest what the writer was thinking at the time.

Examples of mathematical proof

The following proofs illustrate:

1. proof by cases;

2. proof by contrapositive;

3. direct proof; and

4. proof by contradiction.

Proposition 1. For all integers n, $n^3 + 2n$ is exactly divisible by 3.

> [*First thoughts:* Look for a factor 3. Can't find it. Factorise as far as possible: $n^3 + 2n = n(n^2 + 2)$. The proposition would be true if n were a multiple of 3. But what about the other possibilities? Try **proof by cases.**]

Proof: $n^3 + 2n = n(n^2 + 2)$

Case 1: $n = 3k$ for some k.

$$n^3 + 2n = 3k((3k)^2 + 2) \quad \text{which is divisible by 3.}$$

Case 2: $n = 3k + 1$ for some k.

$$n(n^2 + 2) = (3k + 1)((3k + 1)^2 + 2)$$

and we need to look at the second factor.

$$
\begin{aligned}
(3k + 1)^2 + 2 &= 9k^2 + 6k + 1 + 2 \\
&= 3(3k^2 + 2k + 1) \quad \text{(divisible by 3).}
\end{aligned}
$$

Case 3: $n = 3k + 2$ for some k.

Again looking at the second factor, we have

$$
\begin{aligned}
n^2 + 2 &= (3k + 2)^2 + 2 \\
&= 9k^2 + 12k + 4 + 2 \\
&= 9k^2 + 12k + 6 \\
&= 3(3k^2 + 4k + 2) \quad \text{(divisible by 3).}
\end{aligned}
$$

In each case $n^3 + 2n$ is divisible by 3. And the three cases cover all possibilities, since every integer falls into one of the three cases.

So the proposition is true for all integers n.

Proposition 2. If n^2 is even then n is even.

[*First thoughts:* Given n^2 is even we can say $n^2 = 2k$ for some k. Try the **contrapositive**. That way we can use what we are given about n to prove something about n^2.]

Proof: Let us prove 'if n is odd then n^2 is odd'.

Since n is odd we have $n = 2k + 1$ for some k.

Therefore $n^2 = (2k+1)(2k+1) = 4k^2 + 4k + 1$ which is odd.

The proposition in quotes is true and so is the contrapositive of the proposition. Thus, if n^2 is even then n is even.

We observe that this result is quite general. It would be correct to attach the quantifier 'for all integers n' to the proposition.

Proposition 3. If $p > 3$ and p is prime then $p^2 - 1$ is divisible by 3.

[We are given two pieces of information. Perhaps we could start with that and deduce the conclusion, namely $p^2 - 1$ is divisible by 3, by **direct proof**. The strategy is succcessful provided we keep an eye on the conclusion.]

[**Analysis:**   We try to argue forward from what is given and backward from what we have to show, hoping the two arguments will meet.

	given:	to be proved:
statement:	$p > 3$ and p is prime	$p^2 - 1$ is divisible by 3
argument:	$p \neq 2, \quad p \neq 3$	try factors
		$p^2 - 1 = (p-1)(p+1)$

Now observe that $(p-1), p, (p+1)$ are consecutive integers.]

Proof: One of the three numbers $(p-1), p, (p+1)$ is divisible by 3, since they are consecutive.

But that number is not p, since p is a prime greater than 3. Therefore $(p-1)(p+1)$ is divisible by 3.

Therefore $p^2 - 1$ is divisible by 3.

Proposition 4. There are more primes than any natural number. We might say the number of primes is infinite.

The classic proof is due to Euclid (c.300 BC). The method is called **proof by contradiction**.

Proof: Suppose the proposition is false and there is some finite number, k, of primes.

Let the primes be $p_1, p_2, \ldots p_k$, where k is a positive integer. Consider the number $n = p_1 * p_2 * \cdots * p_k + 1$.

n is larger than any known prime, so is not itself prime. That is n must be composite.

Therefore we can express n as a product of primes (proposition 6, p. 99).

But none of the known primes divides n exactly; each leaves remainder 1.

So there must be some prime other than the known primes.

Contradiction. The supposition made at the beginning of the proof must be wrong, so there is not a finite number of primes.

5.2 MATHEMATICAL INDUCTION

Mathematical induction is a style of proof applicable to predicates on the natural numbers. Algorithms with a loop structure are often associated with predicates on a loop counter which is of course a natural number. For this reason alone we would expect mathematical induction to be an important method of proof in computing.

The principle of mathematical induction

In words Let there be a predicate defined on the natural numbers. Suppose that the predicate is true for some (basic) value of the variable. And suppose that if the predicate is true for any value of the variable it is also true for that value plus 1. Then the predicate is true for all values of the variable from the basic value on.

In symbols Let $P(n)$ be a predicate defined on the integer n.
Suppose that $P(a)$ is true for some integer a, and suppose $(P(k) \rightarrow P(k+1))$ is true for all integers $k \geq a$. Then $P(n)$ is true for all $n \geq a$.

It is accepted that the principle of mathematical induction holds for predicates on integers. The principle cannot hold on the real numbers since there is no 'next' number after k when k is real. The principle may be applied to any system in which each variable has a next value, and there is a definite first value.

The following are examples of propositions that may be proved by mathematical induction.

1. The number $n^3 + 2n$ is divisible by 3 for all integers $n \geq 1$.

2. A vending machine is to be programmed to deliver a given value of stamps on request. The machine holds three-cent and five-cent stamps. Provided stocks do not run out, the machine can deliver exactly any value greater than or equal to 8 cents.

3. For all integers $n \geq 1$, $\quad 3^n \geq 1 + 2n$.

4. For all integers $n > 6$, $\quad n! > 3^n$.

5. For all integers $n > 0$,

$$a + ar + ar^2 + \cdots + ar^{n-1} = \frac{a(r^n - 1)}{r - 1} \text{ for } r \neq 1.$$

6. Every integer greater than 1 can be written as a product of primes.

A proof by mathematical induction must prove two propositions.

1. **Basic:** The predicate is true for some *basic* value of the variable.

2. **Inductive:** For all values of the variable from the basic value, if the predicate is true for a value of the variable then the predicate is true for the *next* value of the variable.

The examples below have been chosen to illustrate the method of proof by induction and some of the well-known variations in its application. Proofs by different methods are possible for at least some of the propositions.

Examples of proof by mathematical induction

Proposition 1. The number $n^3 + 2n$ is divisible by 3 for all integers $n \geq 1$.

Define the predicate $P(n)$ to be ($n^3 + 2n$ is divisible by 3).

Proof: Basic: Show that $P(1)$ is true.

$1^3 + 2*1 = 1 + 2 = 3$ and 3 is divisible by 3. So $P(1)$ is true.

Inductive: Show that $(P(k) \rightarrow P(k+1))$ is true for all integers $k \geq 1$.

Assume that $P(k)$ is true. That is, assume $k^3 + 2k$ is divisible by 3.
(*Note*: The use of k is meant to indicate that we are assuming the proposition for only one value of n. At this point in the argument k is constant.) Now we try to show that $n^3 + 2n$ is divisible by 3 for the next value of n, namely $n = k + 1$.

$$(k+1)^3 + 2(k+1) = k^3 + 3k^2 + 3k + 1 + 2k + 2$$
$$= (k^3 + 2k) + 3k^2 + 3k + 3$$

But $k^3 + 2k$ is divisible by 3, by the assumption $P(k)$ is true. Therefore $k^3 + 2k = 3t$ for some t, and

$$(k+1)^3 + 2(k+1) = 3(t + k^2 + k + 1)$$

which is divisible by 3. Therefore $P(k+1)$ is true.

We have shown $P(n) \rightarrow P(n+1)$ is true for a particular constant $n = k$. However the argument is quite general and we deduce that, for every $k \geq 1$, $(P(k) \rightarrow P(k+1))$ is true.

By the principle of mathematical induction $n^3 + 2n$ is divisible by 3 for all $n \geq 1$.

In words, we have shown that whenever the proposition is true for some value of n it is also true for the *next* value of n. But the proposition is true for $n = 1$. Therefore it is true for all values of n from 1 on.

Proposition 2. A vending machine is to be programmed to deliver a given value of stamps on request. The machine holds stocks of three-cent and five-cent stamps. Provided stocks do not run out, the machine can deliver exactly any value greater than or equal to 8 cents.

Define the predicate P(n) to be 'postage of n cents can be made up using three-cent and five-cent stamps'.

Proof: Basic: Show that P(8) is true.

> P(8) is true, since 8 cents can be made up using a three-cent stamp and a five-cent stamp.
>
> **Inductive:** Show that (P(k) → P($k + 1$)) is true for all integers $k \geq 8$.
>
> Assume that P(k) is true. That is k cents can be made up using three-cent stamps and five-cent stamps for some $k \geq 8$.
>
> (*Note*: At this point k is a particular constant value for n.)
>
> Case 1: If there is a five-cent stamp making up part of the k cents, replace it by 2 three-cent stamps to obtain the value $k + 1$ cents.
>
> Case 2: If there is no five-cent stamp in the set making up k cents then there must be at least 3 three-cent stamps to satisfy the condition $k \geq 8$. Replace 3 three-cent stamps by 2 five-cent stamps again making the value $k + 1$ cents.
>
> Thus (P(k) → P($k + 1$)) is true for k, a particular value of n. However the argument holds for every $k \geq 8$.
>
> By the principle of mathematical induction postage of n cents can be made up from three- and five-cent stamps for all values of $n \geq 8$.

Proposition 2 would be hard to prove by direct proof. Further the inductive proof suggests a procedure (perhaps not the most efficient) for programming the vending machine.

Proposition 3. For all integers $n \geq 1$, $3^n \geq 1 + 2n$.

Define the predicate P(n) to be ($3^n \geq 1 + 2n$).

Proof: Basic: Show that P(1) is true.

> $3^1 \geq 1 + 2 * 1$, therefore P(1) is true.

Inductive: Show that $(P(k) \rightarrow P(k+1))$ is true for all integers $k \geq 1$.

Assume that $P(k)$ is true. From this assumption we hope to deduce that $3^{k+1} \geq 1 + 2(k+1)$.

$$
\begin{aligned}
3^{k+1} &= 3 * 3^k \\
&\geq 3 * (1 + 2k) && \text{(assumption } P(k)) \\
&= 1 + 2k + 2 + 4k && \text{(algebra)} \\
&\geq 1 + 2k + 2 && \text{(since } k > 0) \\
\text{Thus} \quad 3^{k+1} &\geq 1 + 2(k+1)
\end{aligned}
$$

That is we have shown $(P(k) \rightarrow P(k+1)$ is true for k a particular value of n.

The argument holds for every $k \geq 1$. By the principle of mathematical induction, proposition 3 is true.

Proposition 4. For all integers $n > 6$, $\quad n! > 3^n$.

Define the predicate $P(n)$ to be $(n! > 3^n)$.

Proof: Basic: Show that $P(7)$ is true.

$$7! = 7 * 6 * 5 * 4 * 3 * 2 * 1 = 5040 \text{ and } 3^7 = 2187$$

Therefore $\quad 7! > 3^7 \quad$ and $P(7)$ is true.

Inductive: Show that $P(k) \rightarrow P(k+1)$ is true for all integers $k \geq 7$.

Assume that $P(k)$ is true. From this assumption we hope to deduce that $(k+1)! > 3^{k+1}$.

$$
\begin{aligned}
(k+1)! &= (k+1)k! \\
&> (k+1)3^k && \text{(assuming } P(k)) \\
&> 3 * 3^k && \text{(since } k \geq 7) \\
\text{Thus} \quad (k+1)! &> 3^{k+1}
\end{aligned}
$$

We have shown $P(k) \rightarrow P(k+1)$ is true for k a particular value of n. However the argument is general and holds for all integers $k \geq 7$.

By the principle of mathematical induction $n! > 3^n$ for all integers $n > 6$.

Proposition 5. For all integers $n > 0$,

$$a + ar + ar^2 + \cdots + ar^{n-1} = \frac{a(r^n - 1)}{r - 1} \text{ for } r \neq 1.$$

Define the predicate P(n) to be (The sum of the first n terms

$$a + ar + ar^2 + \cdots + ar^{n-1} = \frac{a(r^n - 1)}{r - 1}, \quad r \neq 1.)$$

The series part of the predicate uses n in two ways. The formula for the last term is a function of n; and n counts the number of terms in the series. Thus P($n + 1$) will have a new formula for the last term of the series *and* one extra term in the series.

Proof: We are given that $r \neq 1$ so the formula to be proved is defined.

Basic: Show that P(1) is true.

For $n = 1$ the series has just one term, a. The formula for the sum of the series is

$$\frac{a(r^1 - 1)}{r - 1}$$

which is also a. Therefore P(1) is true.

Inductive: Show that (P(k) $\rightarrow$ P($k + 1$)) is true for all integers $k \geq 1$.

Assume P(k) is true. That is, suppose that the formula gives the correct sum for the series with k terms (k constant).

Add the next term in the series, namely ar^k, to both sides of the equation for the sum of k terms that has been assumed. Then the sum of $k + 1$ terms is:

$$
\begin{aligned}
a + ar + &\cdots + ar^{k-1} + ar^k \\
&= \frac{a(r^k - 1)}{r - 1} + ar^k \\
&= \frac{ar^k - a}{r - 1} + \frac{ar^k(r - 1)}{r - 1} \\
&= \frac{ar^k - a + ar^{k+1} - ar^k}{r - 1} \\
&= \frac{a(r^{k+1} - 1)}{r - 1}
\end{aligned}
$$

Thus it is proved that P($k + 1$) follows from the assumption that P(k) is true.

That is $(P(n) \rightarrow P(n+1))$ is true for a particular value $n = k$. However the argument is quite general and holds for all $k \geq 1$. By the principle of mathematical induction the proposition is proved.

Proposition 6 Every integer greater than 1 can be written as a product of primes.

Note: the statement of the proposition contains a slight abuse of the term 'product'. A single prime is taken to be a 'product' of one prime.

Define $P(n)$ to be (each integer from 2 to n can be written as a product of one or more primes).

Proof: Basic step. Show that $P(2)$ is true.
The integer 2 is prime; therefore $P(2)$ is true.
Inductive step. Show that $(P(k) \rightarrow P(k+1))$ is true for all integers $k \geq 2$.
Assume that $P(k)$ is true, (k constant).
Case 1: Suppose $k+1$ is prime. That is $k+1$ is the 'product' of one prime and $P(k+1)$ is true.
Case 2: Suppose $k+1$ is not prime. Then $k+1$ is composite and may be written as a product $m \times n$ where both factors belong to the set $\{2, 3, \ldots, k\}$.
The assumption that $P(k)$ is true means that both m and n are products of primes.

Therefore the same is true of $k+1$. Therefore $P(k+1)$ is true. We have shown $P(k) \rightarrow P(k+1)$ is true for a particular constant k. Now this argument is quite general and therefore $(P(k) \rightarrow P(k+1))$ is true for each $k \geq 2$.

Therefore, by the principle of mathematical induction, every integer greater than 1 can be written as a product of primes.

In words we have shown that whenever the proposition is true for some value of n it is also true for the *next largest* value of n. But the proposition is true for $n = 2$. Therefore it is true for all values of n from 2 on.

5.3 INDUCTION & WELL-ORDERING

To make transparent the underlying logic of mathematical reasoning, particularly mathematical induction, it may help to express the argument using 'rules of inference' or 'rules of deduction'.

A **rule of inference** is a logical *schema* using symbols $p, q, \cdots$ which may be replaced by particular propositions to give an instance (i.e. an application) of the rule. Let us consider an example of a rule of inference.

$$\text{The schema} \qquad \frac{\begin{array}{c} p \\ p \to q \end{array}}{q} \qquad \text{is read}$$

'if we know (or accept or are given) that a proposition (p) is true and we know that the proposition (if p then q) is true then we conclude that (q) is true'.

This rule of inference is called *modus ponens*.

Example: The following is an instance of *modus ponens*.

$$\begin{array}{rl}
1. & x \text{ is even and } x \text{ is prime} \\
2. & \text{if } (x \text{ is even and } x \text{ is prime}) \text{ then } (x = 2) \\
\hline
(\text{Therefore}) & x = 2.
\end{array}$$

The schema shows the logical structure of the argument.

Line 1 is proposition p of the schema, in this case a compound proposition that we assume to be true or *given*.

Line 2 is the proposition $p \to q$ where q is the predicate $x = 2$, and the proposition $p \to q$ is assumed true.

The conclusion of *modus ponens* is that q follows and we have $x = 2$.

5.3.1 The principle of mathematical induction: Weak form

Expressed as a rule of inference, the principle of mathematical induction is the schema

$$\frac{\begin{array}{ll} (\text{basic proposition}) & P(0) \\ (\text{inductive proposition}) & \text{for all } k \geq 0, \, (P(k) \to P(k+1)) \end{array}}{\text{for all } n \geq 0, \, P(n)}$$

It is appealing to regard this principle as a 'repeated application' of *modus ponens*.

$$\begin{array}{rll}
\text{Thus } (1) & P(0) & (\text{basic}) \\
(2) & \dfrac{P(0) \to P(1)}{} & (\text{inductive, } k = 0) \\
(3) & P(1) & (\text{lines 1,2, modus ponens}) \\
(4) & \dfrac{P(1) \to P(2)}{} & (\text{inductive, } k = 1) \\
(5) & P(2) & (\text{lines 3,4, modus ponens})
\end{array}$$

and there follows an infinity of applications of modus ponens.

This is plausible. However mathematicians are conservative and prefer to make the infinity of applications a separate assumption and that assumption is the principle of mathematical induction.

The question does arise however, is there a way to 'prove' that mathematical induction is correct or must we accept it as a basic assumption? Put another way, does the word 'principle' mean 'theorem' or 'basic assumption'? The answer to the first question is that we may prove that mathematical induction is correct if we assume that the natural numbers are well-ordered. Equally however we may prove that the natural numbers are well-ordered if we assume that the principle of mathematical induction is correct. The situation is complicated a little by the fact that the principle of mathematical induction comes in several forms. These forms exist to simplify the application of the principle although they are equivalent in logic.

5.3.2   The principle of mathematical induction: Strong form

Expressed as a 'rule of inference' the strong form of mathematical induction is the schema

$$
\begin{array}{ll}
\text{(basic)} & P(0) \\
\text{(inductive)} & \underline{\text{for all } k \geq 0,\, (P(0) \wedge P(1) \wedge \cdots \wedge P(k) \rightarrow P(k+1))} \\
& \text{for all } n \geq 0,\, P(n)
\end{array}
$$

Further variations are possible. The basic proposition could be $P(0) \wedge P(1) \wedge \cdots \wedge P(m)$, in which case the inductive proposition would be reduced to 'for all $k \geq m, (P(k-m) \wedge P(k-m+1) \wedge \cdots \wedge P(k) \rightarrow P(k+1))$'. Another variation is to start the induction at 1 or some other number instead of 0.

Example 1: Every natural number greater than 1 can be factored as a product of primes, where it is understood that a prime is to be regarded as a 'product' of one prime.

This example has been proved by the weak form of mathematical induction, proposition 6 (p. 99). The proof below uses a slightly different predicate and the strong form of mathematical induction.

    Proof: Define $P(n)$ to be the predicate

$$(n \text{ is a product of prime factors})$$

Basic: $P(2)$ is true because 2 is a prime.

Inductive: Prove $(P(2) \wedge P(3) \wedge \cdots P(k) \rightarrow P(k+1))$ is true for constant $k \geq 2$.

Case 1: $k+1$ is prime. Therefore $P(k+1)$ is true.

Case 2: $k+1$ is composite, so $k+1 = mn$ for some positive integers m, n in the sequence

$$2, 3, \cdots k$$

By the strong inductive assumption, $P(m)$ is true and $P(n)$ is true. Therefore:

$$
\begin{aligned}
k+1 &= \text{(a product of primes)} \times \text{(a product of primes)} \\
&= \text{(a product of primes)}
\end{aligned}
$$

Therefore $P(k+1)$ is true for each k greater than or equal to 2. Therefore by the strong form of mathematical induction, every natural number greater than 1 can be factored as a product of primes.

Example 2: The sequence

$$x_0, x_1, x_2, \cdots, x_n, \cdots$$

is defined by the formula $x_n = 3x_{n-1} - 2x_{n-2}$ for all $n \geq 2$ and the initial conditions $x_0 = 0$ and $x_1 = 1$.

Prove that $x_n = 2^n - 1$.

Informal induction The following table shows that the formula is correct for $n = 0, 1, 2, 3, 4$. But this kind of induction is like scientific experiment. There is no guarantee that the formula is correct for all n.

n	x_n	$2^n - 1$	Formula is correct
0	0	$1-1$	T
1	1	$2-1$	T
2	$3 \times 1 - 2 \times 0 = 3$	$4-1$	T
3	$3 \times 3 - 2 \times 1 = 7$	$8-1$	T
4	$3 \times 7 - 2 \times 3 = 15$	$16-1$	T
	$\cdots$		

Proof by mathematical induction Define $P(n)$ to be the predicate

$$(x_n = 2^n - 1)$$

$P(0)$ is true, line 1 of the table above.

$P(1)$ is true, line 2 of the table above.

Therefore the basic proposition $P(0) \wedge P(1)$ is true.

Assume $P(k-1) \wedge P(k)$ is true for a constant $k \geq 1$.

Then $x_{k-1} = 2^{k-1} - 1$ and $x_k = 2^k - 1$.

$$
\begin{aligned}
\text{Therefore} \quad x_{k+1} &= 3x_k - 2x_{k-1} \quad \text{(given)} \\
&= 3(2^k - 1) - 2(2^{k-1} - 1) \quad \text{(assumption)} \\
&= 3 * 2^k - 3 - 2 * 2^{k-1} + 2 \\
&= 3 * 2^k - 2^k - 1 \\
&= 2 * 2^k - 1 \\
&= 2^{k+1} - 1
\end{aligned}
$$

Therefore $P(k+1)$ follows, and the inductive proposition holds for any $k \geq 1$.

By the strong form of mathematical induction the formula is correct for all natural numbers n. We have proved that $x_n = 2^n - 1$ for all $n \geq 0$.

It is possible to show that any proposition that can be proved by strong induction can be proved by weak induction, using a modified predicate. Also any proposition that can be proved by weak induction can be proved by strong induction since the assumptions needed in the strong form include those of the weak form. In the next section the equivalence of the strong and weak form of mathematical induction is proved; it is also shown that if a set is well-ordered then mathematical induction is a satisfactory method of proof on that set; and that if mathematical induction is valid then the set must be well-ordered.

5.3.3 Well-ordering

In the following theorem it is shown that the well-ordering principle guarantees that each form of mathematical induction is correct and that any one of these three 'principles' may be taken as a basic assumption from which the other two may be proved.

Theorem 5.1 *If a proposition can be proved under the assumption of any one of the three principles given below, then it may be proved using either of the other principles.*

  1. *The strong form of the principle of mathematical induction.*

2. *The weak form of the principle of mathematical induction.*

3. *The natural numbers satisfy the well-ordering principle*
   (definition 3.15, p. 59).

The structure of the proof of this theorem is to show that:
   (1) if we assume the weak form we can deduce the strong form;
   (2) if we assume the strong form we can prove the well-ordering property;
   (3) if we assume the well-ordering property we can prove the weak form.

Proof (1): Assume the weak form.

Strong mathematical induction follows schema (1) below:

$$
\begin{array}{ll}
1. & P(0) \\
2. & P(0) \wedge P(1) \wedge \cdots P(k) \to P(k+1) \text{ for constant } k \geq 1 \\
\hline
\text{Therefore} & P(n) \text{ for all } n \geq 1
\end{array}
$$

Define the new predicate $S(k) = P(0) \wedge P(1) \wedge \cdots P(k)$
Then $S(0) = P(0)$ and schema (1) above can be replaced by

$$
\begin{array}{ll}
1. & S(0) \\
2. & S(k) \to S(k+1) \text{ for constant } k \geq 0 \\
\hline
\text{Therefore} & S(n) \text{ for all } n
\end{array}
$$

and this is a schema for weak induction, given to be true in this proof.

Proof (2): Assume the strong form of mathematical induction.

We aim to prove that the natural numbers are well-ordered.

The well-ordering principle for natural numbers states 'every non-empty subset of **N** has a smallest member'. We prove the contrapositive 'if A is a subset of **N** with no smallest element then A is the empty set'.

Let A be a subset of **N** with no least element.

Let $P(n)$ be the predicate (n belongs to A', the complement of A).

(Basic). The number 0 is not in A, otherwise it is the least element of A.

Therefore 0 belongs to A' and $P(0)$ is true.

(Inductive). Suppose $P(0) \wedge P(1) \wedge \cdots P(k)$ is true.

Then $0, 1, \cdots k$ belong to A', the complement of A.

Now consider $k + 1$. If $k + 1$ is in A then it is the least element of A.

But A has no least element. Therefore $k + 1$ is not in A.

Therefore $P(k+1)$ is true, and by the strong form of mathematical induction it follows that $P(n)$ is true for all natural numbers.

Therefore all natural numbers belong to A' and A is the empty set. Therefore the natural numbers are well-ordered.

Proof (3): Assume: 1. the natural numbers are well-ordered;

2. $P(0)$ is true;

3. $P(k) \to P(k+1)$ for all $k \geq 0$.

We aim to prove that $P(n)$ is true for all $n \geq 0$.

Let F be the set of natural numbers t on which the predicate $P(t)$ is false. We hope to show that F is in fact the empty set.

$$F = \{t \in \mathbf{N} \mid P(t) = \mathsf{F}\}$$

If $F \neq \emptyset$ then F has a least element m by the well-ordering principle and $P(m)$ is false.

Now $m \neq 0$ because $P(0)$ is true (2).

Therefore $m \geq 1$ and $m-1$ is a natural number.

Now $m-1$ is not in F since it is less than the least element of F. Therefore $P(m-1)$ is true, and by (3) $P(m-1) \to P((m-1)+1)$ is true.

Therefore $P(m)$ is true.

The contradiction resulted from the assumption $F \neq \emptyset$.

Therefore F is the empty set and the proposition is proved.

That completes the proof that the three principles are equivalent statements about the natural numbers.

5.4 EXERCISES

For questions 1 to 6 there are some hints. Try the questions first. If you can't see what to do after two or three minutes look at the hints placed after question 6.

1. Prove by cases.

   For all integers n, $n(n+1)(n+2)$, is divisible by 3.

2. Prove by taking the contrapositive.

   For all integers m, n, if mn is odd then m is odd and n is odd.

3. Prove by a direct method.

   For all integers $n \geq 0$, $n^3 - n$ is exactly divisible by 3.

4. Prove by contradiction.

   For all positive integers a, b, n, if $ab = n$ and $a < \sqrt{n}$, then $b > \sqrt{n}$.

5. Prove by mathematical induction.

   For all integers $n \geq 3, \quad n^2 > 2n + 1$.

6. Prove that the following algorithm is correct.

   The algorithm interchanges the values of a and b.

input:	Integer a, b.
	Precondition $P \equiv (a = X, b = Y)$.
	Note: X, Y are just the particular values given to a, b, respectively, at the beginning of the algorithm.
output:	Integer a, b.
	Postcondition $Q \equiv (a = Y, b = X)$.
method:	1. $temp \leftarrow b$
	2. $b \leftarrow a$
	3. $a \leftarrow temp$

Hints:

1. Case $i : n = 3k + i - 1$ for $i = 1, 2, 3$.

2. Prove 'if m is even or n is even then mn is even'.

3. Factorise $n^3 - n$.

4. Assume the proposition false; assume '$ab = n$ and $a < \sqrt{n}$ and $b \leq \sqrt{n}$ '.

5. $P(n) \equiv (n^2 > 2n + 1)$. $P(3)$ is T. Show $P(k) \rightarrow P(k+1)$ for all $k \geq 3$.

6. Insert assertions after lines 1, 2 that give the value of each variable in terms of X and Y. The values of X and Y do not change during the execution of the algorithm.

7. Let $p(n)$ be the predicate $(5^n \geq 1 + 4n)$.

   (a) Write down $p(1)$. Is $p(1)$ true?

   (b) Write down $p(k)$.

   (c) Write down $p(k+1)$.

   (d) Prove that for all $k \geq 1$, if $p(k)$ is true then so is $p(k+1)$.

8. Let the integer n be represented by the string of digits abc in base 10.

   (a) Give the set of possible values for each of a, b, c.

   (b) Give the literal expansion for n.

   (c) Prove that if $b = a + c$ then n is divisible by 11.

(d) Is the converse of the proposition in (c) true?

(e) Construct a more general proposition from (c).

9. Prove that for all integers $n \geq 4$, $\quad n! > 2^n$.

10. Prove that for all integers $n \geq 5$, $\quad 2^n > n^2$.

11. The sum of the first n terms of an arithmetic series whose first term is a and whose common difference is d is defined to be S_n.

$$S_n = a + (a + d) + (a + 2d) + \cdots + (a + (n - 1)d)$$

Prove that $S_n = \frac{n}{2}[2a + (n - 1)d]$.

12. The sequence S_n is defined by

$$S_n = \begin{cases} 2S_{n-1} + 1 & \text{for } n \geq 1, \\ 0 & \text{for } n = 0. \end{cases}$$

Prove that $S_n = 2^n - 1$.

13. Prove for all natural numbers n, $\quad n^2 - 2$ is not divisible by 3.

14. Let $P(n)$ be the predicate ($n^2 + n + 1$ is even).

Prove that $P(k) \rightarrow P(k + 1)$ for all positive integers k. For what (if any) values of n is $P(n)$ true?

15. Prove for all integers $n \geq 0$,

$$2(-1)^n + (-1)^{n+1} = (-1)^n$$

16. Prove by mathematical induction that $n^2 > n+1$ for all integers $n \geq 2$.

17. Prove by mathematical induction that any postage greater than or equal to two cents can be made up exactly using only 2- and 3-cent stamps.

18. A vending machine is being designed to dispense 2-cent and 5-cent stamps only. Assuming sufficiently large stocks of each stamp, prove that any postage from 5 cents or more may be provided by the machine.

19. Let the integer n be represented by the string of digits abc in base 10. Prove that if $a + b + c$ is divisible by 9 then n is divisible by 9.

20. Papers by Mel Friske ('Teaching Proofs', *The American Mathematical Monthly*, **92**, 2, p. 142 (February 1985)) and Uri Leron ('Structuring Mathematical Proofs', *The American Mathematical Monthly*, **90**, 3, p. 174 (March 1983)) examine ways to improve our ability to prove.

Read these papers.

Chapter 6

Recursion

> Recursion is a powerful problem solving tool and a useful pro-
> gramming technique. It is unfortunate that recursive methods
> are not used as much as they should be. (Grogono & Nelson
> (1982, p.236).)

Recursion is important in computing. It may be used as a tool for thinking about a problem, for constructing an algorithm, for constructing a definition or as an explicit programming tool. It can be a way to have the machine, rather than the programmer, sort out difficulties. In this chapter we look at recursion for problem solving and for the construction of algorithms and definitions. The structures to which recursive thinking is applied in this chapter are sequences, lists and functions. You should learn to use recursive formulas, to solve recurrence relations and, at a higher level, to develop your ability to use recursive thinking in problem solving.

Some problems can be arranged so that the solution of a simpler problem of the same kind would solve the initial problem. If (possibly) repeated applications of this strategy of simplification leads to a problem that *can* be solved then the original problem is solved. Such an approach is called *recursive*. The approach may also be applied to the construction of some algorithms and the construction of some definitions. Recursion requires:

- a *base* case for which the solution of the problem, the output for the algorithm or the definition is known;

- a *recursive step* which gives the solution to the general problem, algorithm or definition in terms of a 'simpler' version of the problem, algorithm or definition. 'Simpler' means closer to the base case.

Part of the mathematical importance of recursion is that propositions involving recursion lend themselves to a definite procedure for proof (that is, by mathematical induction).

6.1 RECURSIVE THINKING

6.1.1 Recursive definition

Up to this point we have written expressions such as

1. $1 + 2 + \cdots + 99$

2. $20 \times 19 \times \cdots \times 2 \times 1$

as though it were known what was intended. The '$\cdots$' in the above is supposed to mean that the pattern of the first few terms is continued until the last term. But the pattern is not stated explicitly and hence it would not be clear to an 'algorithm constructor' what should be done. A (nice) constructive way to write an expression containing a repetitive procedure is by means of a *recursively* defined function.

We define functions *sum* and *factorial* on the natural numbers as follows.

1. sum

$$sum : \mathbf{N} \to \mathbf{N}$$

$$sum(n) = \begin{cases} 0 & \text{if } n = 0 \\ n + sum(n-1) & \text{if } n \neq 0 \end{cases}$$

Example 1: Calculate $sum(99)$.

Solution: $sum(99) = 99 + \underbrace{sum(98)}$

To evaluate $sum(98)$ apply the definition again
$$98 + \underbrace{sum(97)}$$

$$\vdots$$

The recursive calls stop at $n = 0$ with the evaluation of $sum(0) = 0$.

The deferred additions are then evaluated, giving:

$$(99 + (98 + \cdots (1 + 0) \cdots)) = 4950.$$

2. factorial

$$factorial : \mathbf{N} \to \mathbf{N}$$

$$factorial(n) = \begin{cases} 1 & \text{if } n = 0 \\ n \times factorial(n-1) & \text{if } n \geq 0 \end{cases}$$

The usual notation for *factorial(n)* is $n!$ With this notation $n! = 1$ if $n = 0$ or $n \times (n-1)!$, if $n \geq 1$.

Example 2: Calculate *factorial*(20), (that is, calculate 20!).

   Solution: $factorial(20) = 20 * \underbrace{factorial(19)}$

To evaluate *factorial*(19) apply the definition again
$$19 * \underbrace{factorial(18)}$$

$$\vdots$$

The recursive calls stop at $n = 0$ and $factorial(0) = 1$.

The deferred multiplications are then evaluated, giving:

$$(20 \times (19 \times \cdots(1 \times 1)\cdots)) = 2.4329 \times 10^{18}.$$

Sigma or summation notation

The *summation* or *Sigma* (Greek letter Σ) notation is used to indicate repeated addition as in example 1.

If f is a function defined on the natural numbers, then the summation notation is defined for natural numbers n, a, where $n \geq a$, by:

$$\sum_{j=a}^{n} f(j) = f(a) + f(a + 1) + \cdots + f(n).$$

The sum on the right has $n - a + 1$ terms. (The variable j is called a dummy variable since it may be substituted by another variable without affecting the sum.)

For example:

$$\sum_{j=1}^{99} j = 1 + 2 + \cdots + 99$$

The notation may be used to express the literal expansion of a number in base x as in Chapter 2. Thus:

$$n = a_k a_{k-1}...a_0 = \sum_{j=0}^{k} a_j x^j$$

To avoid a definition involving '$\cdots$', the summation notation may be defined recursively as follows:

$$\sum_{j=a}^{n} f(j) = \begin{cases} 0 & \text{if } n < a \\ f(n) + \sum_{j=a}^{n-1} f(j) & \text{if } n \geq a \end{cases}$$

Example 1: Calculate:

$$\sum_{i=6}^{8}(2i - 3)$$

$$\sum_{i=6}^{8}(2i - 3) \;=\; (2 \times 8 - 3) + \sum_{i=6}^{7}(2i - 3)$$

$$(2 \times 7 - 3) + \sum_{i=6}^{6}(2i - 3)$$

$$(2 \times 6 - 3) + \sum_{i=6}^{5}(2i - 3)$$

$$0$$

$$= \;(2 \times 8 - 3 + (2 \times 7 - 3 + (2 \times 6 - 3 + 0)))$$

$$= \;33$$

Some general results about Σ notation are useful.

$$\sum_{j=a}^{n}(f(j) + g(j)) \;=\; \sum_{j=a}^{n} f(j) + \sum_{j=a}^{n} g(j) \qquad (6.1)$$

$$\sum_{j=a}^{n} c \;=\; c(n - a + 1) \quad (c \text{ is constant}) \qquad (6.2)$$

$$\sum_{j=a}^{n} cf(j) \;=\; c\sum_{j=a}^{n} f(j) \qquad (6.3)$$

Example 2:

$$\sum_{j=6}^{8} 2j - 3 \;=\; \sum_{j=6}^{8} 2j + \sum_{j=6}^{8}(-3) \quad (\text{by } 6.1)$$

$$= \; 2\sum_{j=6}^{8} j + (-3)(8 - 6 + 1) \quad (\text{by } 6.3, \, 6.2)$$

$$= \; 2(6 + 7 + 8) + (-3)(3)$$

$$= \; 33$$

6.1.2 Problem solving

Solutions of the following examples are intended to illustrate the recursive approach to problem solving.

Example 1: Find a recursive solution for the number of binary n-digit sequences with no consecutive 0's.

A binary sequence is a sequence of 0's and 1's. A 5-digit binary sequence that satisfies the condition is 11010. The sequence 00010 does not satisfy the condition.

Define a_n to be the number of n-digit binary sequences with no consecutive 0's.

For $n = 1$ the possible sequences are 0, 1. Thus $a_1 = 2$.

For $n = 2$ the possible sequences are 11, 10, 01. Thus $a_2 = 3$.

Now every sequence of length n ends in either 1 or 0. And there is exactly one sequence of length n ending in 1 for each legitimate sequence of length $n - 1$, since placing a 1 at the end of a sequence does not break the required condition. For a sequence with more than one member to end in 0 it is necessary that it end in 10, to meet the required condition. And there is exactly one sequence of length n ending in 10 for each legitimate sequence of length $n - 2$. Therefore:

$$
\begin{aligned}
a_n &= \text{number ending in } 1 + \text{number ending in } 10 \\
&= a_{n-1} \quad + \quad a_{n-2}
\end{aligned}
$$

The recursive solution to the problem is then:

$$
a_n = \begin{cases} 2 & \text{if } n = 1 \\ 3 & \text{if } n = 2 \\ a_{n-1} + a_{n-2} & \text{if } n \geq 3 \end{cases}
$$

The recursive formula generates the sequence

$$2, 3, 5, 8, 13, \cdots$$

where each term after the second is obtained by adding its two predecessors.

Example 2: Find the number of n-digit ternary sequences that have an even number of 0's. A ternary sequence is a sequence of 0's, 1's and 2's. The sequence 001010 is an example of a ternary sequence that satisfies the condition in the question.

A sequence of length n must end in 0, 1 or 2. Those sequences ending in 1 or 2 are formed by placing a 1 or a 2 after a sequence of length $n - 1$ that satisfies the condition. On the other hand, a sequence ending in 0 is formed by placing a 0 after a sequence of length $n-1$ that has an odd number of 0's.

Now the total number of sequences of length $n-1$ is 3^{n-1}. So there are $3^{n-1} - s_{n-1}$ sequences of length $n-1$ with an odd number of 0's. Thus:

$$
\begin{aligned}
s_n &= \text{no. ending in 1 or 2} + \text{no. ending in 0} \\
&= 2s_{n-1} \quad + \quad (3^{n-1} - s_{n-1}) \\
&= s_{n-1} + 3^{n-1}
\end{aligned}
$$

There are two sequences of length one, namely 1, 2. (No 0's is an even number of 0's.) The recursive solution to Example 2 is then:

$$
s_n = \begin{cases} 2 & \text{if } n = 1 \\ s_{n-1} + 3^{n-1} & \text{if } n \geq 2 \end{cases}
$$

The recursive formula generates the sequence

$$2, 2+3, 2+3+3^2, \cdots$$

that is $$2, 5, 14, 41, \cdots$$

Example 3: A loan of \$20 000 is to be paid out in 12 equal annual instalments. The interest rate is 14% per year, compounded yearly. Find the annual repayments.

Total repayments must cover the debt plus the interest. But the interest component of each repayment is different. The interest component reduces each year since the outstanding debt decreases each year.

To calculate the interest bill in a given year it is necessary to know the debt for that year, but we are only given the debt at the beginning(\$20 000) and the requirement that the debt at the end be \$0. However if the debt from the preceding year were known, it would be possible to calculate the interest and, by subtracting the fixed repayment, determine the debt at the end of the year. And the debt at the beginning of the first year is known. So we have the requirements for a recursive solution to finding the debt at the end of each year.

The time-line below illustrates what is happening.

The first year starts at time 0 and ends at time 1. In general the jth year ends at time j.

The debt at the end of the jth year after all payments for the year have been made is represented by \$$d_j$.

The annual payment is represented by \$$r$.

Debt(in $)

$$\text{debt at end of year} \quad = \quad \text{debt last year} + \text{interest} - \text{repayment}$$
$$d_j \quad = \quad d_{j-1} + (.14)d_{j-1} - r$$
$$= \quad (1.14)d_{j-1} - r$$

The **recursive formula** for d_j in terms of d_{j-1} is $d_j = (1.14)d_{j-1} - r$. The **base fact** is $d_0 = 20\,000$. And it is known that $d_{12} = 0$. Repeated applications of the recursive step yield:

$$d_{12} \quad = \quad (1.14)d_{11} - r$$
$$= \quad (1.14)((1.14)d_{10} - r) - r$$
$$= \quad (1.14)^2 d_{10} - r[1 + 1.14]$$
$$= \quad (1.14)^3 d_9 - r[1 + 1.14 + (1.14)^2]$$
$$\vdots$$
$$= \quad (1.14)^{12} d_0 - r[1 + 1.14 + (1.14)^2 + \cdots + (1.14)^{11}]$$
$$= \quad 20\,000(1.14)^{12} - r[1 + 1.14 + (1.14)^2 + \cdots + (1.14)^{11}]$$

Given the fact that the debt is to be paid off in 12 years, we have $d_{12} = 0$ and the equation may now be solved for r. Using the formula for the sum of a geometric series we obtain:

$$r\frac{(1.14^{12} - 1)}{1.14 - 1} \quad = \quad 20\,000(1.14)^{12}$$
$$\text{hence} \quad r \quad = \quad \frac{20\,000(1.14)^{12}(.14)}{1.14^{12} - 1}$$
$$= \quad 3533.3865$$

The repayment would be $3533.39 per year.

Discussion: In example 3 the *base* case, $d_0 = 20\,000$, and the *recursive formula* or *recurrence relation*, $d_j = (1.14)d_{j-1} - r$, are sufficient to generate the sequence $d_0, d_1, \cdots$, but the terms are expressed in terms of r. To determine the value of r, the additional condition $d_{12} = 0$ must be used. The values for d_0, d_{12} are called *boundary* values.

Example 4 (binary search): Suppose that $x = \langle x_1, x_2, \cdots x_n \rangle$ is a sequence in which $x_1 \leq x_2 \leq \cdots \leq x_n$. Mathematicians would describe the sequence as 'monotonic increasing', while computer scientists would describe it as a *sorted* list. The problem is to discover whether some number (a) is a member of the sequence. That is, to find a subscript m such that $x_m = a$ or show that a is not in the sequence. It is not required to find *every* occurrence of a, if more than one member of the sequence x equals a.

The algorithm for *binary search* is described recursively. Then it is set out as an iterative procedure in algorithm 6.1. The search is carried out on the *subscripts* of x to yield a message 'not found', if a is not in s, or a subscript m, if $a = x_m$.

Basic step: If the list is empty, report 'a is not in the list'.
Recursive step: If the list is not empty,
         compare the 'middle' element of the list with a.
             if they are equal then a is found.
             if the middle element is less than a
         search the 'upper' list.
             if the middle element is greater than a
         search the 'lower' list.

In each case, if a is not found the recursive step produces a 'smaller' list that is closer to the empty list. The algorithm cannot discard a list with a in it. So it will either find an 'a' or terminate with a report that there is no a in the list.

Given the increasing sequence $< x_p, x_{p+1}, \cdots, x_m, \cdots x_q >$ from x define the 'middle' term to be the term for which

$$m = \left\lfloor \frac{p+q}{2} \right\rfloor \quad \text{if } p \le q; \quad m \text{ is not defined for } p > q.$$

Of the terms not yet rejected in the search, p is the subscript of the smallest term and q is the subscript of the largest term.

Define the 'upper' list to be the increasing sequence $< x_{m+1}, \cdots x_q >$.

Define the 'lower' list to be the increasing sequence $< x_p, x_{p+1}, \cdots, x_{m-1} >$.

Algorithm 6.1 (Binary search) *To search an increasing sequence to find the location of a particular element a, or to show that a is not in the sequence.*

input: a, the element we are searching for.

A sequence $< x_1, \cdots x_n >$, with the subscripts 1, n of the first and last elements of the sequence.

Precondition: $x_1 \leq x_2 \leq \cdots \leq x_n$.

output: A report.

method: $p \leftarrow 1$

$q \leftarrow n$

Repeat

$$\left\{ \begin{array}{l} m \leftarrow \lfloor \frac{p+q}{2} \rfloor \\ \text{if } a < x_m \\ \quad \text{then } q \leftarrow m - 1 \text{ (search 'lower' list)} \\ \quad \text{else } p \leftarrow m + 1 \text{ (search 'upper' list)} \end{array} \right.$$

until $x_m = a$ or $p > q$.

If $x_m = a$, then report '$x_i = a$';

else report 'a was not found'.

Proof: The proof is by mathematical induction on the number of terms in the sequence.

Define l to be $q - p + 1$ for the sequence $< x_p, ...x_q >$. Thus l is the number of terms in the sequence.

Define $P(l)$ to be the predicate 'the algorithm correctly searches all increasing sequences $< x_p...x_q >$ with l terms'.

Basic: $l = 1$. Show that $P(1)$ is true.

If $l = 1$ then $p = q$ and the sequence contains one term, x_p. The subscript m is computed to be $\lfloor \frac{p+p}{2} \rfloor = p$. If $x_p = a$ that fact is reported. If $x_p > a$ the value of q is reduced ($q \leftarrow m - 1$) and we exit the loop with $p > q$. The algorithm terminates with the correct report that 'a was not found'. If $x_p < a$ the value of p is increased ($p \leftarrow m + 1$) and we exit the loop with $p > q$. Again the algorithm terminates with the correct report that 'a was not found'. Hence $P(1)$ is true.

Inductive: Show $P(1) \wedge P(2) \cdots \wedge P(k) \rightarrow P(k + 1)$.

Assume that the algorithm searches correctly all increasing sequences with k terms or less where $k \geq 1$. Consider an increasing sequence with $k + 1$ terms. There will be at least two terms in the sequence. The effect of the application of the algorithm is to either find a or to remove at least one of the terms of the sequence (x_m must go). The result is a sequence that will be searched correctly (by the inductive assumption).

By the (strong) principle of mathematical induction the algorithm is correct for all increasing sequences with one or more terms.

The 'inductive step' corresponds to one pass through the 'repeat' loop. Each such pass either finds a or 'halves' the number of terms in the sequence to be searched. At worst $\log_2 n$ 'halvings' would be required to reduce the number of terms in the sequence to one. For example a sequence with 1000 terms would require ten halvings to reduce to one term. The question of efficiency of algorithms is considered in Chapter 7.

Example: Determine if 2.77 is in the sorted sequence

$$\ln 10, \ln 11, \ln 12, \ln 13, \ln 14, \ln 15, \ln 16, \ln 17$$

where each member of the sequence is given correct to two decimal places. We pretend that we know nothing of the function 'ln' other than that values of it are available on a calculator.

sequence $< x_p...x_q >$	$p > q$	p	q	m	x_m	$x_m = a$?
$x_1, \cdots x_8$	F	1	8	4	2.56	no
$x_5, \cdots x_8$	F	5	8	6	2.71	no
x_7, x_8	F	7	8	7	2.77	yes

The search shows that the 7th term in the sequence is 2.77.

6.2 RECURSION & LISTS

A set of elements in a definite order may be called a list. Lists are important in computing. In ordinary language a sentence may be regarded as a list of words; and a word may be regarded as a list of alphabetic characters. Thus a list may consist of a list of elements which are themselves lists. We define a number of functions on lists.

List The set of all lists under consideration is denoted by *List*. Lists may be represented by l_1, l_2, l or written out element by element.

The lists $< a, b, c >, < a, a, b, c >, < c, a, b >$ are different lists.

Element The set of all elements under consideration for a particular list, might be *Char*, the set of characters, or **N**, the set of natural numbers. The set E is used to denote a general set of elements from which the elements of any particular list may be chosen.

Empty list The list with no elements is called the empty list, written $<>$.

Length The number of elements in a list is called its length.

If $l = < x_1, x_2, \cdots x_n >$ then $length(l) = |l| = n$.

Three functions defined on lists are of particular significance. The functions *head, tail, append* provide a base from which all the common list operations may be obtained (by recursion).

Head The head of a list is defined to be the first element in the list. The head of the list l is written $head(l)$ and is an element, not a list. The head is not defined for the empty list.

Tail The tail of a list is the list obtained by deleting the head from the original list. The tail of the list l is written $tail(l)$ and is a list. Tail is not defined for the empty list.

Append The list obtained by placing the sequence l_2 after the sequence l_1 is called $append(l_1, l_2)$ and is written $l_1 \| l_2$.

For example:

(a) $head(< s, e, t, t, e, r >) = s$

(b) $tail(< s, e, t, t, e, r >) = < e, t, t, e, r >$

(c) $append(< c, a, r >, < p, e, t >)$

$$= < c, a, r > \| < p, e, t > = < c, a, r, p, e, t >$$

There are many functions that may be defined on lists. Often it is advantageous to present these definitions in recursive form, by defining a function on a list in terms of the same function on a smaller list. The following examples show how some functions on lists may be defined recursively.

Length Construct a recursive definition of the function 'length' of a list.

$$length : list \to \mathbf{N}$$

$$length(l) = \begin{cases} 0 & \text{if } l = <> \\ 1 + length(tail(l)) & \text{otherwise} \end{cases}$$

Reverse Construct a recursive definition of a function that reverses the order of the elements in a list.

$$reverse : List \to List$$

$$reverse(l) = \begin{cases} l, & \text{if } l = <> \\ reverse(tail(l)) \| < head(l) >, & \text{otherwise} \end{cases}$$

To illustrate the operation of the recursive definition for *reverse* it is applied to the list $< c, a, t >$.

$$
\begin{aligned}
reverse(< c, a, t >) &= reverse(< a, t >) || < c > \\
&= (reverse(< t >) || < a >) || < c > \\
&= ((reverse(<>) || < t >) || < a >) || < c > \\
&= <> || < t > || < a > || < c > \\
&= < t, a, c >
\end{aligned}
$$

Insert Construct a recursive definition of a function that inserts a given character in a given position in a given list.

$$insert : List \times E \times \mathbf{N} \rightarrow List$$

A precondition for insert is $n \leq |l|$. Read $insert(l, e, n)$ as 'insert the character e in position n in the list l'.

$$
insert(l, e, n) = \begin{cases} < e > || l & \text{if } n = 0 \\ < head(l) > || insert(tail(l), e, n - 1) & \text{otherwise} \end{cases}
$$

6.3 RECURSION & SEQUENCES

Sequences may be defined recursively as in examples 1, 2 and 3 in section 6.1. The geometric sequence and the arithmetic sequence are well-known sequences that are defined most naturally by recursion.

Geometric sequence

$$t_n = r t_{n-1} \text{ for } n \geq 2, \quad t_1 = a$$

Unfolding this definition we obtain the geometric sequence

$$a, ar, ar^2, \cdots ar^{n-1}$$

Arithmetic sequence

$$t_n = t_{n-1} + d \text{ for } n \geq 2, \quad t_1 = a$$

Unfolding this definition gives the arithmetic sequence

$$a, a + d, a + 2d, \cdots a + (n - 1)d$$

The recursive definition for a sequence consists of a 'recurrence relation' together with 'initial conditions'.

A *recurrence relation* for a sequence $t_1, t_2, \cdots$ relates each term of the sequence to one or more preceding terms.

The *initial conditions* for that recurrence relation specify the values of sufficient of the terms $t_1, t_2, \cdots$ to apply the recurrence relation.

The number of preceding terms demanded by the recurrence relation equals the number of initial conditions and is called the *order* of the recurrence relation.

Remark (for those who have done calculus): A recurrence relation may be treated as a discrete version of a differential equation.

The following examples illustrate recursive definition of a sequence.

Example 1: The definition

$$a_n = 2a_{n-1} - a_{n-2}, \quad n \geq 2 \quad \text{(recurrence relation)}$$
$$a_0 = 1, a_1 = 2 \quad \text{(initial conditions)}$$

gives the sequence $1, 2, 3, 4, \cdots$ The recurrence relation is of order 2.

Example 2: The definition

$$s_n - s_{n-1} = n - 1, \quad n \geq 2 \quad \text{(recurrence relation)}$$
$$s_1 = 0 \quad \text{(initial condition)}$$

gives the sequence $0, 1, 3, 6, \cdots$ The recurrence relation is of order 1.

Example 3: The recurrence relation

$$a_n = 4a_{n-2}, \quad n \geq 2$$

is of order 2, although there is only one term in the formula for a_n. To determine the terms of the sequence two initial values are required.

Example 4: The definition

$$x_n = \begin{cases} 2x_{n-1} + 1 & \text{if } n \geq 2 \quad \text{(recurrence relation)} \\ 1 & \text{if } n = 1 \quad \text{(initial condition)} \end{cases}$$

gives the sequence $1, 3, 7, 15, \cdots$ This is a first order recurrence.

Example 5: The definition

$$s_n - 2s_{n-1} - 3s_{n-2} = 0, \quad n > 1 \quad \text{(recurrence relation)}$$
$$s_0 = 3, \quad s_1 = 1 \quad \text{(initial conditions)}$$

gives the sequence $3, 1, 11, 25, \cdots$ This is a second order recurrence relation.

The recurrence relations in examples 1, 3 and 5 are described as *homogeneous* with constant coefficients because all the terms are of the same kind, namely terms of the sequence with constant coefficients.

6.3.1   Calculation of the terms of a sequence

Let the sequence be $s_0, s_1, \cdots$.

An *iterative* calculation finds the nth term of the sequence by calculating all the terms $s_0, s_1, \cdots$ up to s_n.

A *recursive* calculation finds s_n by finding the terms on which s_n depends, then finding the preceding terms on which those terms depend, and so on, until the initial conditions are reached; then all the terms required to evaluate s_n are calculated.

A third method for evaluation of s_n is to find an explicit formula for s_n as an expression in n. This is called *solving* the recurrence relation for s_n.

The efficiency of each of these methods varies greatly. Although the efficiency of algorithms is discussed in general in the next chapter, it is worth looking at the problem now in the context of calculating the terms of a sequence.

Consider a sequence $s_0, s_1, \cdots$ defined by a second order recurrence relation. An iterative approach evaluates s_2 from the given values for s_0, s_1; s_3 from s_1, s_2; $\cdots$ and finally s_n from s_{n-1}, s_{n-2}.

That is $n - 1$ evaluations in all.

A recursive approach evaluates s_n from two predecessors (s_{n-2}, s_{n-1}); and each of these from two predecessors; and so on, doubling the number of calls to the function at each stage down to s_0, s_1 whose values are known. To complete the calculation, something between 2^{n-1} and 2^{n-2} function evaluations are required.

To find the 11th term in a second order sequence, an iterative method would require 10 function evaluations and a recursive method would require between 512 and 1024 evaluations. The advantage for the iterative method is overwhelming for anything beyond the first few terms of this type of sequence. The comparison is not universally in favour of iteration however. Often recursion can be arranged to compete with iteration for efficiency.

On the other hand, if it is possible to find the solution of the recurrence relation, the evaluation by formula may be even more efficient than either iteration or recursion. For example it is shown below that the formula for s_n in example 1 is $s_n = n + 1$, and this requires a single calculation by comparison with the iterative method which uses $n - 1$ such calculations.

The iterative scheme for the evaluation of terms in a sequence is easily written as an algorithm (and is easily programmed). The following example illustrates how this may be done.

Example 6: Develop an iterative algorithm to evaluate the recurrence relation in example 1.

The term before a given term may be called its first predecessor and the term before that may be called the second predecessor. The recurrence relation in example 1 may now be stated as 'the current term is twice the first predecessor minus the second predecessor'.

input:	n, a natural number.	
	A precondition is $n > 1$.	
output:	a.	
	A postcondition: (a_n satisfies its definition (Ex.1)).	
method:	$p0 \leftarrow 1$ (*initial value for second predecessor*)	
	$p1 \leftarrow 2$ (*initial value for first predecessor*)	
	for $j = 2$ to n	

$$\begin{cases} a & \leftarrow & 2*p1 - p0 \\ & & (\textit{derived from } a_n = 2a_{n-1} - a_{n-2}) \\ p0 & \leftarrow & p1 \quad (\textit{updating the second predecessor}) \\ p1 & \leftarrow & a \quad (\textit{updating the first predecessor}) \end{cases}$$

6.3.2 Solving a recurrence relation by observing a pattern

A recurrence relation may sometimes be solved by observing some terms in the sequence and guessing a formula for the general term. It is necessary to prove that this formula is correct and this may be done by the method of mathematical induction.

Calculation of a few terms of example 1 (see p. 121) yields:

$$\begin{array}{lll} n = 2 & a_2 = 2*2 - 1 & = 3 \\ n = 3 & a_3 = 2*3 - 2 & = 4 \\ n = 4 & a_4 = 2*4 - 3 & = 5 \end{array}$$

...

We may conjecture that $a_n = n + 1$. But is this correct?

Prove that $a_n = n + 1$ is a correct formula for the general term of the sequence defined in example 1, namely:

$$a_n = 2a_{n-1} - a_{n-2}, \quad n \geq 2. \quad a_0 = 1, a_1 = 2$$

Proof Basic step: $a_0 = 1$ and $a_1 = 2$

Therefore $a_n = n + 1$ for $n = 0$ and for $n = 1$.

Inductive step: Assume that the formula is correct for two specific successive terms of the sequence. That is, assume for a specific k

$$
\begin{aligned}
a_{k-1} &= (k-1) + 1 &= k \\
a_{k-2} &= (k-2) + 1 &= k - 1
\end{aligned}
$$

Then the successor is

$$
\begin{aligned}
a_k &= 2a_{k-1} - a_{k-2} & \text{by the recurrence relation} \\
&= 2(k) - (k-1) & \text{by the assumption} \\
&= k + 1
\end{aligned}
$$

and the formula is correct for $n = k$.

By the principle of mathematical induction:

$$a_n = n + 1 \text{ for all integers } n \geq 0.$$

Examples 2 and 4 (see p. 121) may be solved similarly.

Example 2: $s_n = s_{n-1} + (n-1), n \geq 2. \, s_1 = 0.$
Sequence: $0, 0+1, 0+1+2, 0+1+2+3, \cdots$
The conjecture, $s_n = \frac{n(n-1)}{2}$, may be proved by mathematical induction.

Example 4: $x_n = 2x_{n-1} + 1, n \geq 2. \, x_1 = 1.$
Sequence: $1, 3, 7, 15, \cdots$ The conjecture $x_n = 2^n - 1$ may be proved by mathematical induction.

6.3.3 Second order, linear, homogeneous recurrence relations

It is not possible to guess solutions in general, but there are good methods for finding solutions for particular types of recurrence relations. We will look at second order, homogeneous recurrence relations with constant coefficients.

Theorem 6.1 (Linearity) *If $s_n + as_{n-1} + bs_{n-2} = 0$ has solutions $f(n)$ and $g(n)$, then $c_1 f(n) + c_2 g(n)$ is also a solution, for arbitrary constants c_1, c_2.*

Proof Substitute $c_1 f(n) + c_2 g(n)$ for s_n in the left-hand side of the recurrence relation.

$$
\begin{aligned}
& (c_1 f(n) + c_2 g(n)) + a(c_1 f(n-1) + c_2 g(n-1)) + \\
& \quad b(c_1 f(n-2) + c_2 g(n-2)) \\
= \; & c_1 (f(n) + a f(n-1) + b f(n-2)) + c_2 (g(n) + a g(n-1) + \\
& \quad b g(n-2)) \\
= \; & c_1 \times 0 + c_2 \times 0 \\
= \; & 0
\end{aligned}
$$

Equations for which this theorem holds are said to be *linear*, so we may now state that homogeneous recurrence relations with constant coefficients are linear recurrence relations. The proof does not depend on the order of the recurrence relation, so we conclude that *all homogeneous recurrence relations with constant coefficients are linear*.

Example 7: Find the solution for example 5 repeated below.

$$ s_n = 2s_{n-1} + 3s_{n-2}, \quad s_0 = 3, s_1 = 1. $$

It has been found that solutions to this type of equation have the form

$$ s_n = x^n \quad \text{for some } x. $$

Let us try the solution $f(n) = x^n$ where x is not yet known.

$$
\begin{aligned}
s_n - 2s_{n-1} - 3s_{n-2} &= 0 & (6.4) \\
x^n - 2x^{n-1} - 3x^{n-2} &= 0 & (6.5)
\end{aligned}
$$

Divide through by x^{n-2}:

$$
\begin{aligned}
x^2 - 2x - 3 &= 0 & (6.6) \\
(x - 3)(x + 1) &= 0 & (6.7)
\end{aligned}
$$

Therefore $x = 3$ or $x = -1$.

Solutions to the recurrence relation are $f(n) = 3^n$ and $g(n) = (-1)^n$.

By the linearity theorem:

$$ s_n = c_1 3^n + c_2 (-1)^n $$

and this solution can be shown to be the *general* solution to the recurrence relation 6.4. Note that it has *two* arbitrary constants c_1, c_2 corresponding to the fact that the recurrence is *second* order.

The initial conditions enable us to determine c_1, c_2.

$$
\begin{aligned}
s_0 = c_1 3^0 + c_2 (-1)^0 &= 3 \\
s_1 = c_1 3^1 + c_2 (-1)^1 &= 1 \\
c_1 + c_2 &= 3 \\
3c_1 - c_2 &= 1 \\
\text{Adding} \qquad\qquad 4c_1 &= 4 \\
c_1 &= 1 \\
c_2 &= 2
\end{aligned}
$$

The solution is therefore:

$$ s_n = 3^n + 2(-1)^n \text{ for } n \geq 0. $$

Equation (6.6)

$$ x^2 - 2x - 3 = 0 $$

is called the *characteristic* equation for $s_n - 2s_{n-1} - 3s_{n-2} = 0$.

Example 8 (Fibonacci): The Fibonacci numbers are defined as the terms of the sequence

$$
F(n) = \begin{cases}
1 & \text{if } n = 0 \text{ or } n = 1 \\
F(n-1) + F(n-2) & \text{if } n \geq 2
\end{cases}
$$

a second order homogeneous recurrence relation.
The characteristic equation is

$$ x^2 - x - 1 = 0 $$

with solution:

$$ x = \tfrac{1+\sqrt{5}}{2} \text{ or } \tfrac{1-\sqrt{5}}{2} $$

So:

$$ F(n) = c_1 \left(\tfrac{1+\sqrt{5}}{2} \right)^n + c_2 \left(\tfrac{1-\sqrt{5}}{2} \right)^n $$

Let us write F_n for $F(n)$. From the initial values $F_0 = F_1 = 1$ we obtain:

$$ c_1 = \frac{1+\sqrt{5}}{2\sqrt{5}} \text{ and } c_2 = \frac{1-\sqrt{5}}{2\sqrt{5}} $$

$$ \text{so } F_n = \frac{1}{\sqrt{5}} \left(\frac{1+\sqrt{5}}{2} \right)^{n+1} + \frac{1}{\sqrt{5}} \left(\frac{1-\sqrt{5}}{2} \right)^{n+1} $$

By considering the absolute value of the second term we see that F_n is the integer closest to

$$ \frac{1}{\sqrt{5}} \left(\frac{1+\sqrt{5}}{2} \right)^{n+1} $$

Historical comment: Leonardo of Pisa, called Fibonacci, was an Italian merchant. He travelled to many parts of the Islamic world of his day. His *Liber Abaci* appeared as a manuscript in 1202 and circulated widely. It undoubtedly played a part in introducing Arabic mathematics to Europe. The book contains a problem leading to the sequence $1, 1, 2, 3, 5, 8, \cdots$ where each term after the second is the sum of its two predecessors. The numbers are called Fibonacci numbers, and have appeared in many investigations. In the next chapter they appear in a problem about an algorithm to find the greatest common divisor of pairs of integers.

6.4 EXERCISES

In general, the characteristic equation for a *second* order homogeneous recurrence relation with constant coefficients will be a quadratic equation and as such will have two distinct roots, one repeated root or no real roots. The 'repeated roots' case is considered in exercises 24, 25 below. The 'no real roots' case is given in exercise 31 and may be attempted by those who have knowledge of complex numbers.

1. Compute: (a) $\displaystyle\sum_{i=0}^{3}(i^2 - 1)$ (b) $\displaystyle\sum_{j=2}^{2}(2j + 3)$ (c) $\displaystyle\sum_{i=2}^{1}(i + 1)$

2. A function *product* is defined on the natural numbers by:

$$product(n) = \prod_{i=a}^{n} f(i) = \begin{cases} f(a) \times f(a + 1) \times \cdots \times f(n) & \text{for } n \geq a \\ 1 & n < a \end{cases}$$

Give a recursive definition for the function *product*.

3. (a) Compute $\displaystyle\prod_{i=0}^{5} i^2$ (b) Show that $\displaystyle\prod_{i=0}^{n} i^2 = (n!)^2$ (c) Compute $\displaystyle\prod_{i=3}^{1} 2i$

4. A sequence of squares may be coloured so that each square is red or white. Let a_n be the number of ways of colouring the sequence so that no two red squares are adjacent. Find a recurrence relation for a_n.

5. Strings are to be made from the letters of the alphabet $\Sigma = \{a, b\}$. Let s_n be the number of strings containing n letters that do not contain the sequence ab.

(a) Find a recursive formula for s_n.

(b) Find a formula for s_n in terms of n.

6. Find a recursive equation for s_n the number of n-digit sequences composed of the digits 0, 1, 2, in which no sequence contains consecutive 1's. (For $n = 5$; 00121, 10121 are examples of possible sequences; 20110 is forbidden.)

7. A loan of $30 000 is to be repaid by equal monthly instalments of $500. The interest rate is 15% per annum, compounded monthly. The *outstanding debt* is found by calculating the debt plus interest less repayments at any given time. Let d_n be the outstanding debt at the end of the nth month just after the interest for the month has been calculated and the repayment for the month has been made.

   (a) Write down a recursive formula from which d_n may be calculated.

   (b) Construct an algorithm to determine the number of payments, the total interest paid and the amount of the last payment.

8. A sum of $5000 is deposited into an account that pays 1% per month compounded monthly. From then on, for eighteen months, $200 is added to the account at the end of each month. Let s_n be the savings, including interest, in the account at the end of month n, immediately after payment and interest have been added.

   (a) Find a recurrence relation for s_n.

   (b) Find s_{18}.

9. We seek a key from the following sequence of keys arranged in ascending order.
   $$x_1, x_2, \cdots, x_{100}$$
   Suppose that the key we seek is (unknown to us) actually x_{18}, the eighteenth key in the sequence. Using binary search to examine the keys, which keys would be examined before the right one was found?

10. Each number in the following sequence is stored correct to two decimal places in a list.
    $$\sqrt{150}, \sqrt{151}, \cdots, \sqrt{165}.$$
    Walk through a binary search to determine whether 12.49 is in the list, or show that it is not.

11. The function *remove* maps a character (x) and a list of characters (l) to a list. It is defined recursively by:
    $$remove(x,l) = \begin{cases} l & \text{if } l = <> \\ remove(x, tail(l)) & \text{if } head(l) = x \\ <head(l)> \| remove(x, tail(l)) & \text{otherwise} \end{cases}$$

It is to be understood that the three cases should be applied in order as given in the definition.

Find the output for $remove(e, < e, v, e, n >)$, and describe in ordinary words what 'remove' does.

12. Define, recursively, a function *removefirst* that removes the first occurrence of a character in a list of characters.

13. Define, recursively, a function *replace* that replaces every instance of one character by another in a list of characters.

14. Construct a recursive definition of a function *IsEven* that maps the numbers in the set $\{0, 1, 2, \cdots\}$ to $B = \{T, F\}$, returning T if the number is even and F if the number is odd.

15. Construct a recursive definition for the *length* of a string from an alphabet A. (Refer to page 61 for a discussion of strings.)

16. Construct a recursive definition for the function *power*. The function is defined for integers a, n where $a > 0, n \geq 0$, by:

$$power\,(a, n) = \begin{cases} 1, & \text{if } n = 0, \\ a * a * \cdots * a, (n \text{ factors}) & \text{if } n \geq 1 \end{cases}$$

17. Messages are transmitted over a communication channel using two signals. The transmission of one signal requires one millisecond and the other signal requires two milliseconds. Let x_n be the number of different messages consisting of sequences of these two signals that can be sent in n milliseconds. Each signal in a message follows the previous one immediately. Find a recurrence relation for x_n, including the initial conditions.

18. The sequence $\{x_n\}$ is given by:

$$x_n = \begin{cases} 1 & \text{if } n = 1 \\ 3 & \text{if } n = 2 \\ 3x_{n-1} - 2x_{n-2} & \text{if } n > 2 \end{cases}$$

(a) Construct an algorithm to find x_{200}.

(b) Solve the recurrence relation to find a formula for x_n in terms of n.

19. Find a formula for s_n in terms of n, given $s_0 = 5$ and $s_n = 3s_{n-1}$ for $n \geq 1$.

20. Find a formula for x_n in terms of n given $x_0 = 1$ and $x_n = 3x_{n-1} + 2$ for $n \geq 1$.

21. Find a formula for s_n in terms of n, given $s_0 = 4, s_1 = 1$ and for $n > 1$, $s_n + s_{n-1} - 2s_{n-2} = 0$.

22. Find a formula for a_n in terms of n, given that $a_0 = 4, a_1 = 14$ and $a_n - 6a_{n-1} + 8a_{n-2} = 0$ for $n \geq 2$.

23. Solve the recurrence relation (i.e. find a formula for a_n in terms of n):

$$a_n = a_{n-1} + 6a_{n-2} \text{ if } n \geq 2; \quad a_0 = 2, a_1 = 6.$$

24. Show that $s_n = n5^n$ is a solution of the recurrence relation:

$$s_n - 10s_{n-1} + 25s_{n-2} = 0$$

What is the general solution?

25. Solve the recurrence relation:

$$a_n + 6a_{n-1} + 9a_{n-2} = 0 \quad \text{if } n \geq 2; \quad a_0 = 3, a_1 = 6$$

26. A sequence $\{a_n\}$ is defined recursively:

$$a_0 = 1, a_1 = 3 \text{ and } a_n = 2a_{n-1} - a_{n-2} \text{ for } n \geq 2$$

(a) Calculate a few terms of the sequence and conjecture a formula for a_n as a function of n.

(b) Prove that the formula you found is correct. (Use mathematical induction.)

27. The sequence $\{a_n\}$ is defined recursively:

$$a_n = 2a_{n-1} - a_{n-2} \text{ for } n \geq 2 \text{ and } a_0 = 1, a_1 = 2$$

(a) Calculate a few terms of the sequence and guess the general formula for a_n.

(b) Prove, by mathematical induction, that the formula for a_n is correct.

28. Find and solve a recurrence relation for the number of binary sequences of length n in which each subsequence of 0's is of even length. (Let s_n be the number of such sequences of length n. Partition the set of sequences of length n into those whose last member is 0 and those whose last entry is 1.)

29. Let s_n be the number of strings of length n containing 0's, 1's and (-1)'s with no two consecutive 1's and no two consecutive (-1)'s. Find a recurrence relation for s_n.

30. The number of comparisons and swaps in the worst case for 'bubble-sort' satisfies the recurrence relation $s_n = s_{n-1} + n - 1$. Find the solution and hence the order of the algorithm.

31. Solve the recurrence relation:

$$\text{for } n \geq 2, \quad s_n + 2s_{n-1} + 2s_{n-2} = 0. \quad s_0 = 1, s_1 = -2$$

32. Consider the proposition

$$(\sum_{i=1}^{n} i(i+1) = \frac{1}{3}n(n+1)(n+2)).$$

(a) Write down the proposition for $n = 1$.

(b) Prove that the proposition holds if $n = 1$.

(c) Write down the proposition for $n = k$.

(d) Write down the proposition for $n = k + 1$.

(e) Prove that if the proposition holds for $n = k$ then it must also hold for $n = k + 1$. Begin your proof as indicated below.

Proof: $\displaystyle\sum_{i=1}^{k+1} i(i+1) =$ (recursive property of $\sum$)

thus   (by assumption)

(f) Why is the proposition true for all $n \geq 1$?

Chapter 7

Analysis of Algorithms

In this chapter we use some of the techniques developed in earlier chapters to construct algorithms, to prove algorithms correct and to measure the efficiency of algorithms. It is shown that there are algorithms so 'slow' that even the fastest computers could not complete them in reasonable time. A measure of performance called the 'complexity' of an algorithm is defined and the ideas of this chapter are illustrated with some algorithms for searching and for sorting.

7.1 SLOW ALGORITHMS

An essential feature of an algorithm is that it should stop in a finite number of steps. However, should this finite number be very large the algorithm may not be practical. There are algorithms which are simple to describe but nevertheless too time-consuming for even the fastest computers to complete in a reasonable time.

In section 7.1 we examine two problems. In each case, a solution algorithm is found using a recursive approach. While an attempt is made to find efficient algorithms, it turns out that each algorithm is 'slow'; as the size of the problem is increased, the number of steps required to execute the algorithm increases quite disproportionately.

In the following section, the measurement of execution properties (time and space requirements) of an algorithm, is considered.

7.1.1 The Towers of Hanoi

The Towers of Hanoi are three spikes on which can be placed 64 disks. Each disk is of a different size and each disk has a hole at its centre so that it can be slid down the spike. At the beginning, all 64 disks are arranged *in order* on the first spike so that the largest disk is at the bottom and the smallest disk is at the top. The other two spikes are empty. The objective is to transfer the disks one at a time from spike to spike so that all disks are transferred to the third spike, arranged in the same order as at the beginning. An algorithm to solve the problem should carry out the transfer in the minimum number of 'disk moves' and so that at no stage is a larger disk on top of a smaller disk.

The problem appears to lend itself to a *recursive approach* in that the problem for 64 disks could be solved if we knew how to solve the problem for 63 disks.

Let the spike that holds all the disks at the beginning be labelled A, let the spike to which all disks are to be transferred be C and let the other spike be B.

Define s_n to be the least number of moves (disk transfers) required to transfer n disks in order on A from A to C according to the rules. The same number of moves are required to transfer n disks in order on one spike to another spike.

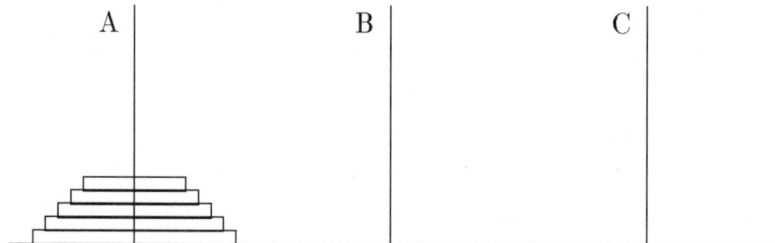

Figure 7.1: Towers of Hanoi for $n = 5$

Consider the Towers of Hanoi problem with n disks on A and the other spikes empty. Transfer $n - 1$ disks from A to B. This transfer requires s_{n-1} moves. Now the largest disk on A, the nth disk, is exposed and may be transferred to spike C in one move. The $n - 1$ disks on B are then transferred to C, taking a further s_{n-1} moves, and completing the transfer. Counting moves, we have:

$$s_n = s_{n-1} + 1 + s_{n-1}$$

that is, the first order recurrence relation:

$$s_n = 2s_{n-1} + 1.$$

For a first order recurrence relation one initial condition is required and we may obtain this by considering one disk. One disk requires one move, therefore $s_1 = 1$.

Applying the recurrence relation successively:

$s_1 = 1, s_2 = 3, s_3 = 7, s_4 = 15, \cdots$

Observing that each of these terms is one less than a power of 2, we may guess

$$s_n = 2^n - 1$$

The conjecture is indeed true and may be proved correct by mathematical induction.

Our algorithm for the Towers of Hanoi requires $s_{64} = 2^{64} - 1$ moves. Now 2^{64} is approximately $1.8 * 10^{19}$. At one millisecond per move the algorithm would take 584 542 000 years.

7.1.2 The travelling salesman

A salesman has to visit a number of towns and seeks to make the best choice of route to minimise the distance travelled. Every pair of towns is joined by a road but the salesman wishes to choose the shortest route which visits each town once only. In the version given here the salesman starts at a given town and visits each other town once only, not returning to the original town. Each possible salesman's route corresponds to a distinct order of the towns. An exact algorithm to solve the problem must generate each route and compare them to find the shortest route. To estimate the amount of work in the algorithm we count the number of routes for a given number of towns.

Let the salesman start at town T_0, and suppose there are n towns, $T_1, T_2, \cdots T_n$, to be visited.

Define s_n to be the number of routes that start at T_0 and visit each of the other towns (n of them) exactly once.

$$
\begin{array}{llll}
n = 1 & \text{There is one route} & T_0T_1 & s_1 = 1 \\
n = 2 & \text{There are two routes} & T_0T_1T_2 & \\
& & T_0T_2T_1 & s_2 = 2
\end{array}
$$

The town T_3 may be introduced into each route with $n = 2$ in three places. For example T_3 may be introduced into $T_0 \triangle T_1 \triangle T_2 \triangle$ in any one of the places marked with a $\triangle$. For each route with $n = 2$ there are three routes with $n = 3$. Therefore, $s_3 = 3s_2$.

In general, $s_n = ns_{n-1}$ for all $n \geq 2$ and $s_1 = 1$. Applying the recurrence relation for successive values of n:

$$s_1 = 1, \quad s_2 = 2 * 1, \quad s_3 = 3 * 2 * 1, \quad \cdots$$

We conjecture (and could prove by mathematical induction) that $s_n = n!$

For $n = 21, s_n = 21!$ which is approximately $5 * 10^{19}$. Supposing that the calculation of the distance for a route with 21 towns takes 1 millisecond, then the time for finding the best route, using our algorithm, would be comparable to that for the Towers of Hanoi algorithm with 64 disks.

The travelling salesman problem has practical value. There are problems with a similar structure, for example choosing a sequence of jobs for a multipurpose machine to minimise clean-up and set-up costs. Because the algorithm given is so time-consuming for anything larger than about $n = 12$, there is a strong motivation to look for faster algorithms. We have a need to develop methods for comparing the execution properties of algorithms.

7.2   COMPARING ALGORITHMS

For a given problem there will be many correct algorithms. It is desirable to have criteria by which alternative algorithms for the problem may be compared and which may be used to improve algorithms. Short execution time, low usage of computer memory space and good properties for 'maintenance' are used as criteria for judging algorithms. Maintenance is concerned with modifying algorithms to meet new requirements. It is thought that good writing, block structure and clear specification will lead to cheaper 'maintenance'. It is also hoped that these features will lead to programs that may be *proven* correct, rather than merely tested for selected inputs.

In this section, however, we consider only the resources of time and space required by algorithms and show how to construct 'complexity' functions to 'measure' the execution time and the space requirement of algorithms. Time and space complexity functions are used:

- to compare different algorithms for the same task; and

- to investigate changes of execution time and space requirement with the size of the task.

The functions used for comparing algorithms are not defined rigorously because they are simplified estimates of the quantity being investigated. Experimental execution times are machine dependent and often accurate estimates are not justified.

Two algorithms are presented to solve the following problem. Time complexity functions are defined for each algorithm and used to decide which algorithm is preferred.

Problem 7.1 *How should we evaluate a degree n polynomial*

$$p(x) = a_n x^n + a_{n-1} x^{n-1} + \cdots + a_1 x + a_0$$

at a particular value of x?

Algorithm 7.1a To evaluate a polynomial of degree n.

> **input:** $x, a_n, a_{n-1}, \cdots a_1, a_0$, real numbers.
> **output:** $p(x)$, real.
> **method:** Multiply out each term and add
> > $p \leftarrow a_0$.
> > For $i = 1$ to n
> > > $t \leftarrow a_i$
> > > for $j = 1$ to i
> > > > $t \leftarrow t * x$
> > > $p \leftarrow p + t$
> > Report that '$p(x)$ is p'.

Consider the number of operations (multiplications and divisions) on the kth pass through the outer loop. For this pass $i = k$. The inner loop is now 'for $j = 1$ to k, $t \leftarrow t*x$' which requires k multiplications. The next (and last) line of the outer loop requires one addition. Thus the number of operations required for $i = k$ is $(k + 1)$. The total number of operations is $(1+1)$ for $i = 1$ plus $(2+1)$ for $i = 2$, ... that is:

$$(1 + 1) + (2 + 1) + \cdots + (n + 1) = 2 + 3 + \cdots + (n + 1) = \frac{n(n + 3)}{2}$$

Algorithm 7.1b To evaluate a polynomial of degree n.

> **input:** $x, a_n, a_{n-1}, \cdots a_1, a_0$, real numbers.
> **output:** $p(x)$, real.
> **method:** Horner's method of nested multiplication.
> > $p \leftarrow a_n$
> > For $i = 1$ to n
> > > $p \leftarrow p * x + a_{n-i}$
> > Report that $p(x)$ is p.

There is one multiplication and one addition on each pass through the loop. The algorithm passes through the loop n times. Therefore the number of operations is $2n$.

The operation count for algorithm 7.1a is $\frac{n(n+3)}{2}$ and for algorithm 7.1b it is $2n$. As n increases, the calculation required by algorithm 7.1a exceeds that of algorithm 7.1b by an ever-increasing amount.

Size of problem n	Algorithm 7.1a $\frac{n(n+3)}{2}$	Algorithm 7.1b $2n$
1	2	2
10	65	20
100	5150	200
200	20 300	400

Assuming that multiplication and addition take about the same time, which is reasonable for modern arithmetic units, the above operation counts give a figure approximately proportional to execution time for the algorithm. This is the intention behind the definition of the 'complexity' function.

On the basis of 'time complexity' the second algorithm would be preferred to the first algorithm for evaluation of any polynomial of degree $n \geq 2$.

Definition 7.1 (Complexity) *The* time complexity *of an algorithm is defined as a function T. The domain variable(s) specify the 'size' of the problem. It is preferred to use a single size variable if possible. The range variable is proportional to execution time but does not equal it due to simplification.*

Algorithm 7.1a has time complexity $T_1 : N \to R, \quad T_1(n) = \frac{n(n+3)}{2}$

Algorithm 7.1b has time complexity $T_2 : N \to R, \quad T_2(n) = 2n$

Definition 7.2 (Space complexity) *The* space complexity *of an algorithm is defined as a function S. The domain variable(s) specify the size of the problem. The range variable is proportional to the requirement for memory space for storage of words, strings or numbers by the algorithm.*

Table 7.1 shows values for some complexity functions at four domain points, namely $n = 5, 10, 100, 1000$.

T	$\log_2 n$	$\sqrt{n}$	n	n^2	n^3	2^n	$n!$
$n = 5$	2.3	2.2	5	25	125	32	120
10	3.3	3.2	10	100	1000	1024	$3.6 * 10^6$
100	6.6	10	100	10 000	10^6	$1.27 * 10^{30}$	–
1000	9.9	31.6	1000	1 000 000	10^9	–	–

Table 7.1: Complexity function values

The symbol '–' indicates that the number is larger than 10^{100}.

In general it is considered that algorithms comparable with 2^n or $n!$ are too slow for practical use. On the other hand, algorithms that behave no worse than n^2 or even n^3 are considered to be sufficiently fast to have practical value.

Consider the expression $T_1(n) = \frac{n(n+3)}{2}$ or $T_1(n) = \frac{1}{2}n^2 + \frac{3}{2}n$ written as a polynomial. Writing in the form

$$T_1(n) = n^2 \left(\frac{1}{2} + \frac{3}{2}\frac{1}{n} \right)$$

and observing that $\frac{1}{n}$ is small for large n, it may be seen that the complexity function could be approximated by $\frac{1}{2}n^2$ for large n.

The coefficient $(\frac{1}{2})$ is relatively unimportant, so that an algorithm is often described by the highest order term in the complexity function, ignoring the numerical coefficient. For example Algorithm 7.1a would be described as 'order n^2' and Algorithm 7.1b as 'order n'.

Example: Suppose that an algorithm with time complexity

$$T(n) = \frac{1}{2}n^2 + \frac{3}{2}n$$

takes 1 second to complete a task for which $n = 500$. How long would it take for a task for which $n = 5000$?

Solution (using 'order'). The algorithm is of order n^2.

$$\begin{aligned}
\text{Time} \quad &= \quad 1 * \frac{5000^2}{500^2} \quad \text{seconds} \\
&= \quad 100 \quad \text{seconds}
\end{aligned}$$

A more accurate estimate using the complexity function T is:

$$\begin{aligned}
\text{Time} \quad &= \quad 1 * \frac{T(5000)}{T(500)} \\
&= \quad 1 * \frac{(5000)^2/2 + 3 * 5000/2}{(500)^2/2 + 3 * 500/2} \\
&= \quad 99.46 \quad \text{seconds}
\end{aligned}$$

In this example the *order* of the algorithm was sufficient to give a good indication of the effect of increasing the size of the task. The order of a function is defined below.

Definition 7.3 (Order) *A function $f(n)$ is said to be of order $g(n)$, written $O(g(n))$ if there exists a constant c such that:*

$$|f(n)| < c|g(n)| \text{ for all } n \geq n_0 \geq 0.$$

The definition does not tell us how to *choose* $g(n)$. The practice is to choose the 'simplest' function for g that will represent the 'behaviour' of f for 'large' n.

Example 1: $f(n) = \frac{1}{2}n^2 + \frac{3}{2}n$ is $O(n^2)$ since $|f(n)| \leq 2|n^2|$ for all $n \geq 0$.

The function whose value at n is $\frac{1}{2}n^2 + \frac{3}{2}n$ is quadratic in n. The simplest representative of quadratic functions is n^2 so $g(n)$ is chosen to have value n^2.

Example 2: $f(n) = 2n + \log_2 n^3$ is of order n.

$$
\begin{aligned}
2n + \log_2 n^3 &= 2n + 3\log_2 n \\
&< 2n + 3n \quad \text{for } n \geq 1
\end{aligned}
$$

Therefore, $|2n + \log_2 n^3| < 5|n|$ for $n \geq 1$

Therefore $f(n)$ is $O(n)$.

The interest is in the behaviour of $f(n)$ for 'large' positive n so it is not important that $n = 0$ is avoided.

Example 3: $f(n) = 2n + n\log_2 n$ is of order $n\log_2 n$.

$$
\begin{aligned}
2n + n\log_2 n &= n(2 + \log_2 n) \\
&< n(\log_2 n + \log_2 n) \quad \text{for } n > 4 \\
&= 2n\log_2 n \quad \text{for } n > 4
\end{aligned}
$$

Therefore, $|f(n)| < 2|n\log_2 n|$ for $n > 4$

If an algorithm has time complexity $T(n)$ then it is said to be of order $O(T(n))$. Any algorithm that is $O(p(n))$, where $p(n)$ is a polynomial, is said to be a *polynomial time algorithm*. If $p(n)$ is a cubic polynomial, we say the algorithm is $O(n^3)$. In general we look for the simplest function $g(n)$ so that if the algorithm is polynomial of degree s we say it is $O(n^s)$. An algorithm whose time complexity function is not polynomial is called an *exponential time algorithm*.

If it is *not* possible to solve a problem in polynomial time, then the problem is said to be *intractable*.

The word 'intractable' has been given a special meaning here. Some complexity functions are not uniformly larger or smaller than others; for example, $2^n > 5n^2$ for $n \geq 8$. But the inequality is reversed for $n < 8$. Also the complexity function is usually defined as a 'worst-case' measure. In practice some algorithms consistently perform faster than the complexity function would suggest.

For example, the *simplex* algorithm for linear programming has been shown to have exponential time complexity, but in practice it is relatively quick. Commercial linear programming problems with hundreds of variables are solved as a matter of routine by computer. This well-known fact encouraged researchers to look for polynomial time algorithms for the linear programming problem and algorithms of this type have now been found.

The examples below construct complexity functions for some algorithms from Chapter 1.

Example 1: Algorithm 1.3 (p. 8) tests whether an integer n greater than 1 is prime. In the worst case, the algorithm checks n for divisibility by each integer from 2 to $\lfloor \sqrt{n} \rfloor$. Taking the 'divisibility check' as the unit of time taken by the algorithm, the time complexity is

$$T(n) = \lfloor \sqrt{n} \rfloor - 1$$

The order of the algorithm is $\sqrt{n}$.

Example 2: Algorithm 1.4 (p. 9) is an algorithm to find all the prime factors of a number n.

The algorithm differs from the previous one in that if a factor is found, the search continues for more factors. We must investigate whether the search for possible factors takes more time if there are many factors or if there are no factors. The worst case for 'many factors' is $n = 2^k$. The worst case for 'no factors' is n prime. For $n = 2^k$ the algorithm passes through the first branch $k - 1$ times, where $k = \log_2 n$. For n prime the algorithm passes through the second branch $\sqrt{n}$ times as we found in Example 1. Although the time taken in branch 1 is not the same as branch 2, each pass through a branch is associated with a divisibility test and, taking the divisibility test as the unit of time the worst case is the maximum of $(\log_2 n, \sqrt{n})$. For $n > 16$, $\sqrt{n}$ is the worst case.

The order of this algorithm is therefore $\sqrt{n}$.

Example 3: Algorithm 1.5 (Euclid, p. 11) is an algorithm to find the greatest common divisor of integers m, n where $m \geq n > 0$.

Let the unit of time be the time taken to do a division. Then it is possible to show that the number of divisions is less than $2 \log_2 n$ and Euclid's algorithm is $O(\log_2 n)$. The result depends on the following proposition.

The remainder in Euclid's algorithm halves every second division.

Proof: From the proof of algorithm 1.4 (p. 9) we have:

$$m = n * q_1 + r_1 \text{ and } 0 \le r_1 < n$$
$$n = r_1 * q_2 + r_2 \text{ and } 0 \le r_2 < r_1$$

Case 1: $r_1 \le \frac{n}{2}$,   then $r_2 < r_1 \le \frac{n}{2}$ so $r_2 < \frac{n}{2}$.

Case 2: $r_1 > \frac{n}{2}$,   then $n = r_1 * q_2 + r_2$ and q_2 must be 1.

Therefore $n = r_1 + r_2$ and $r_2 = n - r_1 < n - \frac{n}{2} = \frac{n}{2}$.

In each case we have $r_2 < \frac{n}{2}$.

Subsequent divisions repeat the same pattern so that after any two divisions carried out under the algorithm, the remainder will have halved.

Suppose $n < 2^k$. Then after $2k$ divisions there will be k successive halvings of n and the final remainder will be 0. But $k = \log_2 n$. So the number of divisions is at worst $2 \log_2 n$. In this example we have found an 'upper bound' to the worst case and the algorithm has time complexity

$$T(n) < 2 \log_2 n.$$

The order of the algorithm is $\log_2 n$.

The algorithm given by Euclid around 300 BC is thus a very fast algorithm.

7.3   APPLICATIONS

In the following sections of this chapter we investigate searching and sorting algorithms to illustrate the ideas on algorithm efficiency, algorithm correctness and algorithm development that have been discussed in this book.

7.3.1   Searching

Searching through a sequence of numbers or a sequence of words to find a particular number or word is a frequent computer task. Often the number is a 'key' to a record containing information about a particular individual. For example, each time you 'logon' to a mainframe computer a search must be made using your usercode as a key to check that the correct password corresponding to your usercode has been given.

Problem 7.2 (Search) *How may we find a given item (a) in a list (l) or show that a is not in the list?*

Let the list be $l = \langle x_1, x_2, \cdots, x_n \rangle$. Then we are asked to find an i such that $x_i = a$ or show that no such i exists. If it is possible that a occurs more than once in the list then it might be required to find all occurrences or just one. We will consider the case where to find just one occurrence is enough. The searching problem falls into two basic types:

- searching a random sequence or an unsorted list;

- searching a monotonic increasing sequence or a sorted list.

Algorithm 7.2 (Linear search) *To find an i such that $x_i = a$ where x_i belongs to l, or to show that a is not in l.*

Recursive version

In the following $lsearch(l, i, a)$ may be read as 'carry out a linear search on the list l to find a'. A precondition is $i = 1$.

$$lsearch: \quad list \times \mathbf{N} \times E \to \{\text{'not found'}\} \cup \mathbf{N}$$

The postcondition is $x_i = a$ for the exit value of i or $a \notin l$.

$$lsearch(l, i, a) = \begin{cases} \text{'not found'} & \text{if } i > n \\ i & \text{if } head(l) = a \\ lsearch(tail(l), i+1, a) & \text{otherwise} \end{cases}$$

Iterative version

    input: $l = \langle x_1, x_2, \cdots, x_n \rangle$, a list.
               a, the item being sought (*has the same type as x_i*).
    output: Subscript i or 'not found'.
               The postcondition is $x_i = a$ if a is in l;
               otherwise the output should be 'not found'.
    method: Define i, a natural number.
               (*i is the subscript of the item being considered.*
               $1 \leq i \leq n + 1$)
               Define *found*, Boolean
               $i \leftarrow 1$
               found $\leftarrow$ F
               repeat
                    if $x_i = a$
                        then *found* $\leftarrow$ T
                        else $i \leftarrow i + 1$
               until (*found* or $i > n$)
               if *found*
                    then report ' the ith term, x_i, equals a';
                    else report 'not found'.

The work done in this algorithm may be measured by counting the number of comparisons of x_i with a. The size of the problem is measured by n, the number of items in the list. In the worst case the algorithm will pass n times through the loop giving n comparisons.

In the worst case the time complexity is $T(n) = n$.

The average case time complexity would be $T(n) = \frac{n}{2}$.

The order of the algorithm is n.

Binary search

Algorithm 6.1 (p. 115) describes a search procedure to find an element a in an *increasing* sequence (or *sorted* list). The order of the algorithm may be deduced from a proof that the algorithm terminates.

Proof of termination Algorithm 6.1

> The algorithm must terminate. For on each pass through the loop, the value of $l = q - p + 1$, the 'length' of the sequence left to be searched, is approximately halved. In the worst case this value reaches $l = 1$ after $\lfloor \log_2 n \rfloor + 1$ halvings. At this point $p = q = m$ and there is one x_m to be tested. If $x_m = a$ the algorithm terminates. If $x_m \neq a$ the algorithm changes p or q but in each case so that $p > q$ and again the algorithm terminates.

Order From the argument for termination of the algorithm above we see that in the worst case the algorithm passes through the loop

$$\lfloor \log_2 n \rfloor + 1 \quad \text{times.}$$

Taking the unit of time to be the time for such a pass gives the order of the algorithm to be $\log_2 n$.

7.3.2 Sorting

The speed advantage of binary search (order $\log_2 n$) over linear search (order n) is overwhelming. It is normally worthwhile to sort a list if it is to be 'looked up', that is searched, frequently. In this section it is assumed that for each list or sequence under consideration there is a total order defined on the items of the list or the numbers in the sequence. In practice the items may be words which are to be written in 'dictionary order' or integers or reals to be written in numerical order.

Problem 7.3 *How may a list be sorted so that the items are arranged in increasing order?*

Bubblesort

Algorithm 7.3 (Bubblesort) *To rearrange the items in a list so that they are in increasing order.*

Bubblesort operates on a list by finding the largest item in the list and rewriting the list as:

the list with the largest item removed $||$ < largest item > .

The algorithm then operates recursively on 'the list with the largest item removed' placing successive 'largest items' to the end. The recursion stops when the 'list with the largest item removed' has just one element. At this point the deferred 'append' operations can be carried out to yield the original list sorted in increasing order.

In the version of the algorithm given below:

bsort(l)' should be read as 'carry out a bubblesort on the list l.

max(l)' should be read as 'the largest item or number in l'. The function *max* is defined on non-empty lists and maps a list to its largest item.

l - **max(l)** should be read as 'the list remaining when the largest item is removed'. The operation is the function *removefirst* defined in exercise set 6.4, and could be defined

$$l - max(l) = removefirst(max(l), l)$$

$$
\begin{aligned}
bsort: &\quad List \to List \\
\text{precondition:} &\quad l \text{ is not empty} \\
\text{postcondition:} &\quad l \text{ is sorted}
\end{aligned}
$$

$$bsort(l) = \begin{cases} l, & \text{if } |l| = 1 \\ bsort(l - max(l)) \, || < max(l) >, & \text{otherwise} \end{cases}$$

Proof: Termination

If the list is of length 1 then the algorithm enters the first branch and terminates. If the list is not of length 1 then it must contain more than one item, for by the precondition it is not empty. Under this condition the algorithm enters the second branch. The second branch calls the function again but on a list with one less item. Possibly after repeated calls to the function the list on which it operates will be of length 1 and the algorithm will enter the first branch and all deferred operations will be carried out.

Correctness

Define the predicate $P(n)$ to be '*bsort* correctly sorts a list l where $|l| = n$ '.

A list with one item is sorted. For such a list the algorithm selects the first branch and yields the correct output, namely l. Therefore $P(1)$ is true.

Suppose that P is true for $n = k$ where $k \geq 1$. Then for $n = k+1$ the algorithm selects the second branch. Now $|l - max(l)| = k$ since one of the elements has been removed. Therefore $bsort(l - max(l))$ is correct by the assumption of mathematical induction. And assuming that $max(l)$ meets its specification and $l - max(l)$ meets its specification the output from $bsort(l - max(l))$ is a sorted list consisting of all the items of l, except one, the largest. That largest item is then appended to the end of the list, giving a correctly sorted list l of $k+1$ items. Therefore $P(k + 1)$ is true and by the principle of mathematical induction the algorithm is correct for $n \geq 1$. That is the algorithm operates correctly on all non-empty lists.

Order Let $|l| = n$ be the number of items in l. To find the largest item in l, comparisons of that item with the $n - 1$ others must have been made. Regarding a 'comparison' as the unit of work, the algorithm requires $n - 1$ comparisons to find $max(l)$, then $n - 2$ comparisons to find the largest item in the list from which $max(l)$ has been removed, and so on.

The number of comparisons is

$$(n - 1) + (n - 2) + \cdots + 1 = \frac{n(n - 1)}{2}$$

and the algorithm is of order n^2.

Example: Sort the list $< c, d, a, b >$.

$$bsort(< c, d, a, b >)$$
$$bsort(< c, a, b >) || < d >$$
$$bsort(< a, b >) || < c >$$
$$bsort(< a >) || < b >$$
$$< a >$$

Now carry out the deferred appends (last deferred, first done).

$$< a, b >$$
$$< a, b, c >$$
$$< a, b, c, d >$$

In the above version of bubblesort nothing was stated about how max should operate. The algorithm below has more detailed instructions. The process of providing further detail is called the 'refinement' of the algorithm and ultimately leads to a computer program in some computer language.

method: for $k = 1$ to $n - 1$
             for $j = 1$ to $n - k$
                 if $x_j > x_{j+1}$
                     then $exchange(x_j, x_{j+1})$

refinement of 'exchange'
$$\begin{cases} temp & \leftarrow & x_j \\ x_j & \leftarrow & x_{j+1} \\ x_{j+1} & \leftarrow & temp \end{cases}$$

Mergesort

Mergesort is another sorting algorithm. It splits a sequence into two 'half sequences'. If the halves are not sorted then the algorithm is called again to sort the halves. The two sorted half sequences are 'merged' to form a single sorted sequence. The process of 'halving' the sequences reduces the length of the sequence, so that eventually the sequence produced will be of length 1 or 0. Such a sequence is sorted so the merges can now be done and the algorithm can complete. The algorithm depends on a function *merge* which outputs a single sorted list from an input of two sorted lists.

$merge$: $List \times List \rightarrow List$
precondition: x is sorted and y is sorted
postcondition: z is sorted and z contains all the elements of x, y.
$merge(x, y) \quad =$

$$\begin{cases} x & \text{if } y = <> \\ y & \text{if } x = <> \\ < head(x) > || merge(tail(x), y) & \text{if } head(x) \leq head(y) \\ < head(y) > || merge(x, tail(y)) & \text{otherwise} \end{cases}$$

Example: Use the *merge* function to combine the lists $< 2, 4 >, < 3, 5, 7 >$.

$$merge(< 2, 4 >, < 3, 5, 7 >)$$
$$= \ < 2 > ||merge(< 4 >, < 3, 5, 7 >)$$
$$= \ < 2 > || < 3 > ||merge(< 4 >, < 5, 7 >)$$
$$= \ < 2 > || < 3 > || < 4 > ||merge(<>, < 5, 7 >)$$
$$= \ < 2 > || < 3 > || < 4 > || < 5, 7 >$$
$$= \ < 2, 3, 4, 5, 7 >)$$

Time complexity The 'size' of an application of *merge* depends on the length of each of the input lists. The execution time will be proportional to the number of comparisons between heads of lists. In the worst case there will be a comparison preceding the placement of each element, except the last, in the output list z. Let $length(x) = p$ and let $length(y) = q$ then the value of the time complexity function for merge at (p, q) is $p + q - 1$.

Space complexity The space required for this algorithm is that for each of the input lists plus the space for the output list. That is, $2(p + q)$ where p, q are the lengths of the input lists.

Given a list $l =< x_1, x_2, \cdots, x_n >$ define the sublists:

$$lower(l) = l_1 =< x_1, x_2, \cdots x_m >$$

$$upper(l) = l_2 =< x_{m+1}, \cdots, x_n >$$

where $m = \lfloor \frac{n-1}{2} \rfloor$.

Algorithm 7.4 (Mergesort) *To rearrange the items in a list so that they are in increasing order.*

$$mergesort: \quad List \rightarrow List$$
$$mergesort(l) \ = \ \begin{cases} l, & \text{if } |l| = 1 \\ merge(mergesort(l_1), mergesort(l_2)), & \text{otherwise} \end{cases}$$

Example: Find $mergesort(< 10, 8, 9, 11, 7 >)$. Indicate the split between the lower and upper lists by the symbol $|$.

$$mergesort(< 10, 8, 9, 11, 7 >)$$
$$= \ < 10, 8, 9, |11, 7 >$$
$$= \ (< 10, 8, |9 >) \ (< 11, |7 >)$$

$$
\begin{aligned}
&= ((<10,|8>) \quad <9>) \quad (<11><7>)\\
&= ((<10><8>) \quad <9>) \quad (<11><7>)\\
&= \text{lists are length 1 and sorted; start merges}\\
&= (<8,10> \quad <9>) \quad <7,11>\\
&= <8,9,10> \quad <7,11>\\
&= <7,8,9,10,11>
\end{aligned}
$$

Time complexity To simplify matters let us suppose that the number of items in the list to be sorted is a power of 2. Then $|l| = n = 2^k$ and $\log_2 n = k$.

Define s_n to be the number of comparisons required to sort the list l. Then a recurrence relation for s_n is

$$
s_n = \underbrace{s_{\frac{n}{2}} + s_{\frac{n}{2}}} + \underbrace{\left(\frac{n}{2} + \frac{n}{2} - 1 \right)}
$$

1. $s_{\frac{n}{2}}$ is the number of comparisons required to 'mergesort' each half list of length $\frac{n}{2}$.

2. $\frac{n}{2} + \frac{n}{2} - 1$ is the number of comparisons required to 'merge' two lists of length $\frac{n}{2}$.

$$
s_n = 2s_{\frac{n}{2}} + (n-1) \tag{1}
$$

The formula is correct for n a power of 2. Substitute $\frac{n}{2}$ for n and then multiply both sides by 2.

$$
2s_{\frac{n}{2}} = 4s_{\frac{n}{4}} + (n-2) \tag{2}
$$

Continue in the same way until

$$
2^{k-1}s_2 = 2^k s_1 + (n - 2^{k-1}) \tag{k}
$$

Now $2^k = n, 2^{k-1} = \frac{n}{2}$ and $s_1 = 0$. Summing the k equations we obtain after subtracting out equal terms from both sides of the equations:

$$
\begin{aligned}
s_n &= ns_1 + kn - (1 + 2 + 4 + \cdots + 2^{k-1})\\
&= kn - (2^k - 1)\\
&= kn - (n - 1)\\
&= n\log_2 n - (n - 1)
\end{aligned}
$$

The order of the algorithm is $n\log_2 n$.

The time complexity of mergesort is order $n \log_2 n$ while that of bubblesort is n^2. Thus mergesort is much faster than bubblesort.

The space complexity of bubblesort is n since the new list is written over the old list. The space complexity of mergesort is $2n$ since for each 'merge' the input lists are written to a third output list whose length must be the sum of the lengths of the input list. Thus mergesort is preferred for time but not for space.

7.4 EXERCISES

1. How much time does an algorithm take to solve a problem of 'size' n if this algorithm uses $5n^2 + 2^n$ bit operations each requiring 10^{-6} seconds with the following values of n?

   (a) 10   (b) 20   (c) 40   (d) 80

2. For a given problem two algorithms are available. Algorithm 1 has time complexity $n^2 + n$ and algorithm 2 has time complexity $n \log_2 n + 2n$.

   (a) Which is the faster algorithm, and by approximately how much, for a problem in which $n = 1024$?

   (b) If algorithm 2 takes 1 second to complete a task for which $n = 512$, how long will it take for one in which $n = 8192$?

3. Two algorithms for a given task have computational complexity functions T_1, T_2 where:
$$T_1(n) = 3\sqrt{n} \log_2 n$$
$$T_2(n) = 2n - 1000$$

   (a) Which algorithm is preferable for a task in which $n = 5000$?

   (b) If the second algorithm takes 10 seconds when $n = 5000$, how long would the first algorithm take?

   (c) Algorithm 1 takes 1.05 seconds for a task in which $n = 1000$. How long would it take when $n = 100\,000$?

4. The algorithm *mergesort* may be used to sort a list into numeric or alphabetic order. When applied to a list of $n = 2^k$ items *mergesort* requires
$$T(n) = n \log_2 n - (n - 1)$$

'comparisons' to complete, where a 'comparison' is regarded as a measure of time. Calculate $T(512)/T(256)$ to estimate how the time for *mergesort* increases from a sequence of 256 items to a sequence with twice that number.

5. An algorithm to solve a given problem has time complexity

$$T(n) = n \log_2 n - (n - 1)$$

(a) Calculate $T(4096)$.

(b) Given that the algorithm takes 0.8 second for a problem in which $n = 1024$, how long should it take for a problem in which $n = 4096$?

(c) Suppose another algorithm were available to solve the same problem. If the second algorithm had time complexity

$$T_2(n) = \frac{n(n - 1)}{2}$$

which algorithm would be preferred on the basis of time to solve a problem in which $n = 1024$?

6. For n sufficiently large (say $n > 16$)

$$\log_2 n < \sqrt{n} < n < n^2 < 2^n < n!$$

Place $n \log_2 n$ and $n^{\frac{3}{2}}$ in the correct place in the above ordered sequence.

7. Find the order of:　(a) $n^2 + 2n + 1$　　　(d) $n^2 + 2^n$

(b) $n^2 + n \log_2 n$　　　(e) $n! + 2^n$

(c) $n + n \log_2 n^2$

8. The following algorithms are intended to find the maximum (*max*) of the terms in the sequence $a_1, a_2, \cdots a_n$. Find the time complexity of each algorithm. Which of the algorithms should be selected on the basis of speed of operation?

(a) While the sequence has more than one member

$$\begin{cases} \text{partition the sequence into pairs} \\ \text{discard the smaller number in each pair} \end{cases}$$

max ← the remaining number.

(b) $max \leftarrow a_1$

  for $j = 2$ to n

$$\left\{\begin{array}{l} \text{if } a_j > max \\ \quad \text{then} \quad max \leftarrow a_j \end{array}\right.$$

9. A courier has to visit five offices (labelled 1-5) from his base (labelled 0) and return to the base. The entry in row i column j of the matrix below gives the time in minutes to travel from location i to location j, where $i, j = 0, 1, \cdots 5$.

$$\begin{pmatrix} 0 & 40 & 15 & 60 & 15 & 30 \\ 40 & 0 & 25 & 50 & 40 & 25 \\ 15 & 25 & 0 & 25 & 45 & 55 \\ 60 & 50 & 25 & 0 & 10 & 30 \\ 15 & 40 & 45 & 10 & 0 & 20 \\ 30 & 25 & 55 & 30 & 20 & 0 \end{pmatrix}$$

The courier claims that travel time must be 2.5 hours, which he calculated by following the sequence 0-1-2-3-4-5-0. But the previous courier used to do the trip in two hours! How did she do it? The only other information known was that the previous courier always started by going from the base to office 2.

10. A small jobshop receives an order for five products. Transition costs are incurred when the manufacturing process is changed from one product to another. The number in row i column j of the following matrix represents the cost, in hundreds of dollars, of changing from product i to product j.

$$\begin{pmatrix} 0 & 2 & 14 & 7 & 19 \\ 15 & 0 & 8 & 3 & 6 \\ 9 & 4 & 0 & 5 & 13 \\ 16 & 10 & 11 & 0 & 18 \\ 20 & 21 & 17 & 12 & 0 \end{pmatrix}$$

For example the cost of changing production from product 2 to product 3 is $800.

Find the cheapest manufacturing sequence, given that production may commence with any product. What is the cost?

11. The function below finds the greatest common divisor of a pair of integers a, b satisfying the precondition $(0 \le a < b)$:

$$\gcd(a, b) = \left\{\begin{array}{ll} b & \text{if } a = 0 \\ \gcd(r(b, a), a) & \text{if } a \ne 0 \end{array}\right.$$

where $r(b, a)$ is defined to be the remainder when b is divided by a (i.e. $0 \le r < a$ and $b = aq + r$ for some integer q).

(a) Use the function (recursively, as defined) as an algorithm to calculate $\gcd(21, 34)$, $\gcd(20, 205)$, and $\gcd(23, 37)$. How many function calls are required for each application of the algorithm? Try to identify features of the 'worst case' input for the algorithm.

(b) The Fibonacci number F_n is defined (p. 126) by $F_0 = 1$, $F_1 = 1$ and, for $n \ge 2$, $F_n = F_{n-1} + F_{n-2}$. Show that for $n \ge 2$, $\gcd(F_n, F_{n+1}) = \gcd(F_{n-1}, F_n)$.

(c) Show that, if $n \ge 2$, $\gcd(F_n, F_{n+1})$ requires $n + 1$ recursive calls for evaluation.

(d) Show that for applications of the gcd algorithm, the 'worst case' occurs when a, b are successive terms in the Fibonacci sequence.

(e) Given that F_n is the integer closest to

$$\frac{1}{\sqrt{5}} \left(\frac{1 + \sqrt{5}}{2} \right)^{n+1}$$

find the worst case time complexity for the greatest common divisor algorithm.

(f) Given that 832 040 and 1 346 269 are successive Fibonacci numbers, find the number of function calls required by the gcd algorithm to find their greatest common divisor. What is the greatest common divisor of these numbers?

12. Illustrate how the *mergesort* algorithm operates by showing the effect on the following sequence of each recursive call of the algorithm.

$$\cos 50, \cos 100, \cos 200, \cos 300$$

The angles are measured in degrees.

13. Set your calculator to *radian* mode. Illustrate the bubblesort algorithm by sorting the sequence

$$\cos 1, \cos 2, \cos 3, \cos 4, \cos 5.$$

14. (Requires a knowledge of probability) A binary search for an item A is conducted on an ordered list $x_1, x_2, \cdots, x_{15}$. Given

$$\mathrm{Prob}(A = x_i) = \frac{1}{16}, \quad i = 1, 2, \cdots, 15$$

calculate:

(a) the 'worst case' number of comparisons made during the search.

(b) the 'best case' number of comparisons made during the search.

(c) the expected number of comparisons made during the search.

Chapter 8

Graphs and Trees

Many structures in computing may be modelled by graphs. Apart from an introduction to the language of graphs, this chapter looks at the problem of constructing a spanning tree for a connected graph. You should learn the language of graphs and be able to solve problems about graphs. Some of the problems will require you to select a spanning tree algorithm and apply it. The relation 'connected' is shown to be an equivalence relation on the vertices of a graph and equivalence relations make a link between this chapter and Chapter 3. This chapter makes use of matrices. For those not familiar with matrices a short description including matrix multiplication is provided.

An important theme in this chapter is the relation between pictures (geometry) and algebraic structures too complex to be usefully pictured. The pictures are important to generate new ideas and to understand the ideas of others. But to be effective the pictures must be translated into algebraic structures for large problems which will mostly be handled automatically by a computer. Mathematicians have given the word 'graph' a special meaning. The word describes a set of vertices (or nodes), some or all of which may be connected by edges (or arcs). Graphs are used to model many situations, for example rail networks, computer networks, job schedules, data structures and many more.

8.1 GRAPHS

A graph is a discrete structure consisting of a set of vertices, pairs of which may be joined by edges. Figure 8.1 illustrates a graph in which the vertices (sometimes called nodes) are named v_1, v_2, v_3, v_4 and the edges (sometimes called arcs) are named a, b, c, d, e. A vertex is represented pictorially by a small circle, $\circ$, and an arc by a line or curve joining two vertices.

The edge a is said to be *incident* with vertex v_1 and with vertex v_2.

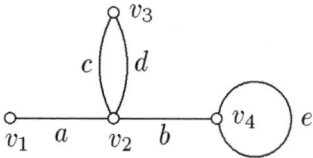

Figure 8.1: A graph

Vertices v_1, v_2 are said to be *adjacent* because they are joined by an edge.
The edges c, d are said to be *multiple* edges because they are incident with
the same pair of vertices v_3, v_2. The edge e is called a *loop* because it is
incident with just one vertex. A graph that does not have multiple edges or
loops is called a *simple* graph. If multiple edges are allowed the structure
is called a *multigraph*. If multiple edges and loops are allowed the structure
is called a *pseudograph*. We will call a graph that allows all possibilities a
general graph. Some authors use different names to those given here but
the names have been chosen to agree with as many writers as possible.

Definition 8.1 (Graph) *A graph G is a set $V(G)$ of objects called ver-*
tices, together with a multiset $E(G)$ of edges. Each element of E is a multiset
of two vertices from V. If $V = \emptyset$ the graph is called the null graph.

A graph $G(V, E, i)$ may be classified using the incidence function

$$i: \quad E \longrightarrow \{\{u, v\} | u, v \in V\}$$

where:

1. if i is one-one and $u \neq v$ then G is a *simple* graph; in this case each edge
   is identified unambiguously by the vertices with which it is incident
   (in this case the multisets are sets);

2. if i is many-one and $u \neq v$ then G is a *multigraph*; in this case multiple
   edges are possible;

3. if i is many-one with no requirement that u, v be distinct then G is
   called a *pseudograph* or *general* graph. Multiple edges and loops are
   both allowed.

Example 1: Suppose $V(G) = \{v_1, v_2, v_3\}, \quad E(G) = \{e_1, e_2\}$ and

$$i(e_1) \mapsto \{v_2, v_1\}, i(e_2) \mapsto \{v_1, v_3\}$$

Then $G(V, E, i)$ is a simple graph with three vertices and two edges. A
picture may be drawn to represent the graph, as in Figure 8.2. Vertices

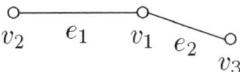

Figure 8.2: Graph for example 1

v_1, v_2 are adjacent but v_2, v_3 are not. The edge e_1 is incident with v_2 and v_1. Since the graph is a simple graph the edges may be described as *2-subsets* of vertices without ambiguity. For example $e_1 = \{v_2, v_1\}$. The edges are pictured as straight lines since that is easy to typeset. However edges may be drawn as curved lines if you wish.

Example 2 (Jugs): Model the following puzzle by a graph and find the solutions.

Three jugs have capacities of 8, 5 and 3 litres but it is not possible to estimate how much is in each if they are part full. Find a sequence of pourings, beginning with the 8 litre jug full and the others empty and ending with 4 litres in each of the 8 litre jug and the 5 litre jug.

Solution: The 'jug system' is supposed to exist in a number of states which may be obtained by a sequence of steps each of which consists of pouring part or all of the contents of one jug into another without spilling. The ordered triple (x, y, z) represents the state in which the 8 litre jug contains x litres, the 5 litre jug contains y litres and the 3 litre jug contains z litres. The system starts in state $(8,0,0)$ and a single pouring takes the system to $(3,5,0)$ or $(5,0,3)$. A single pouring would not produce the state $(4,4,0)$ because it would not be possible to estimate that the 5 litre jug was $\frac{4}{5}$ full. A valid pouring may be defined as a single pouring in which either a jug is filled or a jug is emptied; and the effect may be reversed by a single pouring. The second part of the condition allows us to consider only those pourings that potentially help to solve the problem. The condition excludes such pourings as $(4,4,0)$ to $(8,0,0)$ (which is unhelpful) since the reverse is not possible in a single step.

The 'jug system' is now modelled as a graph in which the vertices correspond to the states of the system and the edges correspond to valid pourings.

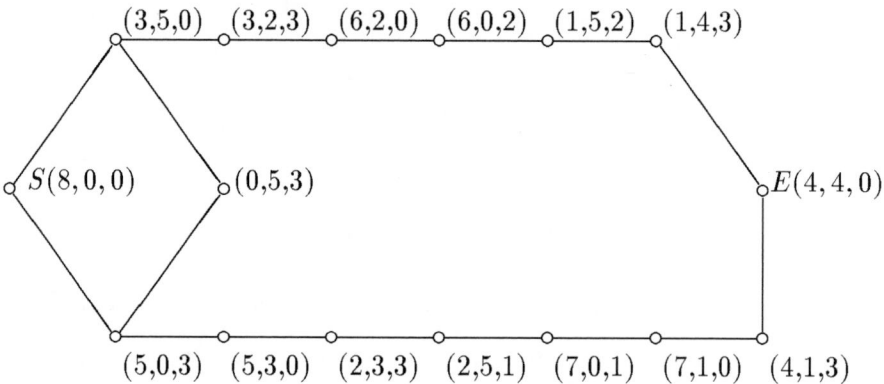

Figure 8.3: Graph for the 'jugs' puzzle

The graph in Figure 8.3 models the 'jugs' puzzle. The graph shows that
there are two main solutions and that once started on a sequence of pourings
the puzzle solver has only to persevere to succeed in finding a solution to
the puzzle.

For each edge drawn it would be possible, although perhaps not useful,
to carry out a pouring in either direction. For example it would be possible
to pour from (3,2,3) to (6,2,0) or from (6,2,0) to (3,2,3). Other pourings
are possible in this system in one direction only. For example it would be
possible to go from state (6,2,0) to (8,0,0) in a single pouring, but not to go
from (8,0,0) to (6,2,0). This fact could be modelled by allowing edges to be
ordered pairs of vertices. Such a directed edge is indicated by an arrow as
shown in Figure 8.4.

Directed edges have been excluded from the 'jugs' model.

$(6, 2, 0)$ $(8, 0, 0)$

Figure 8.4: A directed edge

Definition 8.2 (Digraph) *If the incidence function maps edges to ordered pairs of vertices then the graph is called a* directed graph *or* digraph.
In that case it is usual to place arrows on the edges in a picture of the graph.

Definition 8.3 (Subgraph) *A* subgraph *of a graph G is a graph whose vertices are a subset S of $V(G)$ and whose edges contain only vertices from S and are a subset of $E(G)$.*

Sometimes it is useful to attach numbers to the edges of a graph. For example the numbers might represent distances or flow capacities. Such graphs are said to be *weighted* or *edge labelled* or the graph is called a *network*.

Definition 8.4 (Edge label) *An* edge-labelling *of a graph is a function*

$$elabel : E(G) \rightarrow W$$

where W is a set of labels, usually numbers representing (for example) distances or flows.

Example 1: The graph in Figure 8.5 has a label attached to each edge.

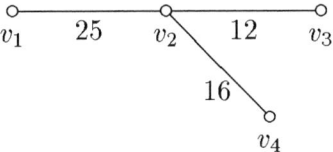

Figure 8.5: An edge-labelled graph

Example 2: The road map for the travelling salesman (Chapter 7) could be modelled by a graph. The towns would be vertices, the roads would be edges and the edge labels might be the road distances.

Definition 8.5 (Vertex label) *A* vertex-labelling *of a graph is a function*

$$vlabel : V(G) \rightarrow W$$

where W is a set of labels for the vertices.

Definition 8.6 (Degree of a vertex) *The* degree *of a vertex is the number of edges incident with the vertex, with each loop counting twice.*

Theorem 8.1 *Let G be a graph with p vertices and q edges. Let vertex v_i have degree d_i for $i = 1, 2, \cdots, p$. Then:*

$$\sum_{i=1}^{p} d_i = 2q$$

Proof: Taking the sum of the degrees of the vertices, each loop is counted twice by definition; and every other edge is counted twice, once for each vertex that it contains. Thus the sum of the degrees of the vertices of a graph is twice the number of edges.

Corollary: The number of vertices of odd degree is even.

8.1.1   Matrices and matrix multiplication

For those not familiar with the subject this section gives a short description of matrices. The following section requires knowledge of matrices and matrix multiplication.

Definition 8.7 (Matrix) *An $m \times k$ matrix A is a rectangular array of elements having m rows and k columns as shown below.*

$$A = \begin{pmatrix} a_{11} & a_{12} & \cdots & a_{1k} \\ a_{21} & a_{22} & \cdots & a_{2k} \\ \cdots & \cdots & a_{ij} & \cdots \\ a_{m1} & a_{m2} & \cdots & a_{mk} \end{pmatrix}$$

The element in the ith row and jth column is a_{ij}.

An example of a 3×4 matrix is:

$$A = \begin{pmatrix} 1 & 2 & 0 & -1 \\ 0 & 3 & 4 & 2 \\ 2 & 2 & -2 & 3 \end{pmatrix}$$

The element a_{23} is 4.

An example of a square matrix is the 3×3 matrix:

$$S = \begin{pmatrix} 1 & 2 & 0 \\ 2 & 4 & 0 \\ 3 & 2 & 1 \end{pmatrix}$$

Definition 8.8 (Matrix multiplication) *If A is an $m \times k$ matrix and B is a $k \times n$ matrix then the product AB exists and is defined to be an $m \times n$ matrix in which the element in row i column j is:*

$$a_{i1}b_{1j} + a_{i2}b_{2j} + \cdots + a_{ik}b_{kj} = \sum_{x=1}^{k} a_{ix}b_{xj}$$

Example 1: A 2×3 matrix multiplied by a 3×2 matrix gives a 2×2 matrix. Thus:

$$\begin{pmatrix} 1 & 2 & -1 \\ 0 & 3 & 2 \end{pmatrix} \begin{pmatrix} 1 & 2 \\ 0 & 3 \\ 2 & 2 \end{pmatrix}$$

$$= \begin{pmatrix} 1 \times 1 + 2 \times 0 + -1 \times 2 & 1 \times 2 + 2 \times 3 + -1 \times 2 \\ 0 \times 1 + 3 \times 0 + 2 \times 2 & 0 \times 2 + 3 \times 3 + 2 \times 2 \end{pmatrix}$$

$$= \begin{pmatrix} -1 & 6 \\ 4 & 13 \end{pmatrix}$$

Observe that changing the order of multiplication would give a different result for the matrices in this example. For some other pairs of matrices it would not be possible to multiply at all in the reverse order.

Example 2:

$$\begin{pmatrix} 1 & 2 \\ 3 & 4 \end{pmatrix} \begin{pmatrix} 0 & 1 \\ 2 & -1 \end{pmatrix} = \begin{pmatrix} 1 \times 0 + 2 \times 2 & 1 \times 1 + 2 \times -1 \\ 3 \times 0 + 4 \times 2 & 3 \times 1 + 4 \times -1 \end{pmatrix}$$

$$= \begin{pmatrix} 4 & -1 \\ 8 & -1 \end{pmatrix}$$

Example 3: Given that

$$A = \begin{pmatrix} 0 & 1 & 1 \\ 1 & 0 & 1 \\ 1 & 1 & 0 \end{pmatrix}$$

calculate A^2, A^3 and A^4.

$$A^2 = AA = \begin{pmatrix} 2 & 1 & 1 \\ 1 & 2 & 1 \\ 1 & 1 & 2 \end{pmatrix}; \quad A^3 = \begin{pmatrix} 2 & 3 & 3 \\ 3 & 2 & 3 \\ 3 & 3 & 2 \end{pmatrix};$$

$$A^4 = \begin{pmatrix} 6 & 5 & 5 \\ 5 & 6 & 5 \\ 5 & 5 & 6 \end{pmatrix}.$$

8.1.2   Description of a graph

Representation of a graph by a picture is helpful if the number of vertices and edges is small. In practice, graphs of commercial value frequently have too many vertices and edges for the picture to be useful. A representation that may be stored in a computer is needed. For the description of the vertices of a graph a list is sufficient. The 'adjacency matrix' and the 'adjacency list' described below are two well-known ways to represent the edges of a graph.

Adjacency matrix The adjacency matrix of a simple graph G with p vertices is the $p \times p$ matrix of 0's and 1's:

$$A = \begin{pmatrix} a_{11} & a_{12} & \cdots & a_{1p} \\ a_{21} & a_{22} & \cdots & a_{2p} \\ \cdots & \cdots & a_{ij} & \cdots \\ a_{p1} & a_{p2} & \cdots & a_{pp} \end{pmatrix}$$

where for all i, j from 1 to p

$$a_{ij} = \begin{cases} 1, & \text{if } v_i \text{ is adjacent to } v_j \\ 0, & \text{if } v_i \text{ is not adjacent to } v_j \end{cases}$$

Note: For each order in which the vertices are considered there will be a different adjacency matrix for a given graph.

Adjacency list For each vertex of a graph, list the vertices adjacent to that vertex. That is, for each v in G define $l(v)$ to be the list of vertices adjacent to v.

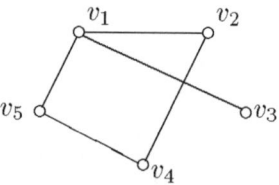

Figure 8.6: A graph

Example 1: Give the 'adjacency matrix' representation for the graph pictured in Figure 8.6.

$$A = \begin{pmatrix} 0 & 1 & 1 & 0 & 1 \\ 1 & 0 & 0 & 1 & 0 \\ 1 & 0 & 0 & 0 & 0 \\ 0 & 1 & 0 & 0 & 1 \\ 1 & 0 & 0 & 1 & 0 \end{pmatrix}$$

The second row, for example, of the matrix is intended to show that v_2 is adjacent to v_1 and v_2 is adjacent to v_4. The other rows give similar information.

Example 2: Give, as a table, the 'adjacency list' representation for the graph pictured in Figure 8.6.

1	2	3	4	5
2	1	1	2	1
3	4		5	4
5				

The first column, for example, of the table is intended to show that v_1 is adjacent to v_2, v_3 and v_5. The other columns give similar information.

The list of vertices adjacent to $v_1, v_2 \cdots$ may be written explicitly as lists. For example, $l(v_1) = < v_2, v_3, v_5 >, \cdots$.

8.1.3   Walks, paths, and other connections

Many problems that may be modelled by graphs require consideration of ways in which vertices may be connected via edges and intermediate vertices in the graph. For example, problems associated with finding shortest or longest distances between vertices and problems associated with touring vertices or edges each require the construction of different kinds of tracks through a graph.

Definition 8.9 (Walk) *A* walk *of length n in a graph G is a sequence of vertices and the edges formed by each successive pair of vertices.*

$$W(v_0, v_n) = v_0, e_1, v_1, e_2, \cdots, e_n, v_n \quad where \ e_i = \{v_{i-1}, v_i\}.$$

In a simple graph the walk may be described by listing the vertices alone, for example $v_0 v_1, \cdots v_n$, where it is understood that the walk contains the edges joining successive vertices.

The *length* of a walk is the number of edges in it.

If $n = 0$ the walk consists of the vertex v_0 and no edge. It is called a walk of length zero. If $v_0 = v_n$ the walk is said to be *closed*.

Definition 8.10 (Path) *A walk with distinct vertices is called a* path.

Definition 8.11 (Trail) *A walk with distinct edges is called a* trail.

Every path is a trail; but not every trail is a path.

Definition 8.12 (Circuit) *A closed trail is called a* circuit.

A circuit is not a path.

Definition 8.13 (Cycle) *A circuit with at least one edge, and in which the only repeated vertex is* $v_0 = v_n$, *is called a* cycle.

A cycle of length n is called an *n-cycle*. A 1-cycle is a loop, a 2-cycle consists of a pair of multiple edges. For $n \geq 3$ an n-cycle may be called a polygon.

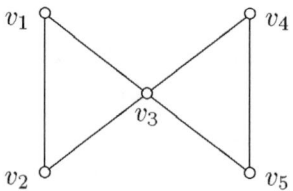

Figure 8.7: A graph with two cycles

Example 1: In the graph of Figure 8.7:

1. $W(v_1, v_5) = v_1 v_2 v_3 v_5 v_4 v_3 v_5$ is a walk of length 6.
2. $W(v_1, v_1) = v_1 v_3 v_1$ is a closed walk of length 2, but it is not a circuit.
3. $P(v_1, v_5) = v_1 v_2 v_3 v_4 v_5$ is a path of length 4.
4. $W(v_1, v_2) = v_1 v_3 v_4 v_5 v_3 v_2$ is a trail but not a path.
5. $W(v_1, v_1) = v_1 v_2 v_3 v_4 v_5 v_3 v_1$ is a circuit but not a cycle.
6. $v_1 v_4 v_3$ is not a walk since $\{v_1 v_4\}$ is not an edge.
7. $W(v_1, v_1) = v_1 v_2 v_3 v_1$ is a 3-cycle.

Walks, paths, trails, circuits and cycles are particular examples of subgraphs of a graph.

The adjacency matrix offers a method for the computation of the number of walks of a given length in a graph. For example, we show below how to calculate the number of walks of length 2 in a given graph.

Example 2: Find, and calculate the number of, all walks of length 2 from v_2 to v_5 in the graph of Figure 8.6.

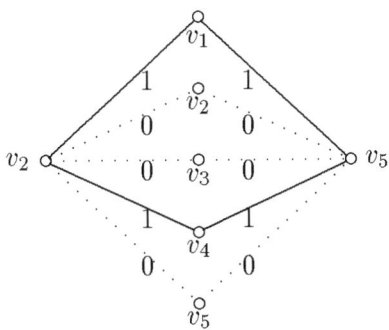

Figure 8.8: Walks of length 2 from v_2 to v_5 in Fig. 8.6

In Figure 8.8, all possible walks of length 2 from v_2 to v_5 in the graph of Figure 8.6 are drawn. Edges which exist are represented by a solid line and are labelled '1' as in the adjacency matrix. Edges which are not in the graph are dotted and labelled '0' as in the adjacency matrix.

Walks of length 2 may be identified by multiplying the labels on the edges that form possible walks to form a walk label. Thus $v_2 v_1 v_5$ has walk label 1×1 which is 1 and the walk is indeed a walk of length 2 from v_2 to v_5. However $v_2 v_3 v_5$ has walk label 0×0 which is 0 and this possible walk does not exist in the graph of Figure 8.6.

In general, let a, b be the edge labels for $v_i v_j$ and $v_j v_k$ respectively. If $ab = 1$ then $v_i v_j v_k$ is a walk and if $ab = 0$ then $v_i v_j v_k$ is not a walk.

The number of all possible walks of length 2 from v_2 to v_5 may be found by the calculation

$$1 \times 1 + 0 \times 0 + 0 \times 0 + 1 \times 1 + 0 \times 0$$

which yields two walks of length 2.

Now the calculation above is exactly the same as a calculation carried out in the normal process of matrix multiplication. For the adjacency matrix A:

$$A^2 = \begin{pmatrix} 0 & 1 & 1 & 0 & 1 \\ 1 & 0 & 0 & 1 & 0 \\ 1 & 0 & 0 & 0 & 0 \\ 0 & 1 & 0 & 0 & 1 \\ 1 & 0 & 0 & 1 & 0 \end{pmatrix} \begin{pmatrix} 0 & 1 & 1 & 0 & 1 \\ 1 & 0 & 0 & 1 & 0 \\ 1 & 0 & 0 & 0 & 0 \\ 0 & 1 & 0 & 0 & 1 \\ 1 & 0 & 0 & 1 & 0 \end{pmatrix}$$

Row 2 of the matrix A on the left represents the edges from v_2 to each other vertex as illustrated on the left-hand side of Figure 8.8. Column

5 of the matrix A on the right represents the edges from each vertex to v_5 as illustrated on the right-hand side of the figure. The normal matrix multiplication yields the result of row 2 times column 5 in position row 2 column 5 of A^2 below. In general the number of walks of length 2 from v_i to v_j is the number in row i column j of A^2.

$$A^2 = \begin{pmatrix} 3 & 0 & 0 & 2 & 0 \\ 0 & 2 & 1 & 0 & \mathbf{2} \\ 0 & 1 & 1 & 0 & 1 \\ 2 & 0 & 0 & 2 & 0 \\ 0 & 2 & 1 & 0 & 2 \end{pmatrix}$$

The above result is generalised in Theorem 8.2 below.

Theorem 8.2 (Counting walks) *Let A be the adjacency matrix of a graph. The element a_{ij}^s of the matrix A^s counts the number of walks of length s from v_i to v_j in the graph.*

Example: Find the number of walks of length 4 from v_1 to v_3 in the graph in Figure 8.9.

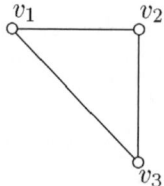

Figure 8.9: Graph for example

Solution: The adjacency matrix for this graph is the matrix A of example 3 on page 161. The entry in row 1 column 3 of A^4 is '5'. By theorem 8.2 above there are five walks of length 4 from v_1 to v_3. (An example of such a walk is $v_1 v_3 v_2 v_1 v_3$.)

8.2   TREES

In general a graph will have more than one path between some pairs of vertices and perhaps no path at all between other pairs. A graph with exactly one path between each pair of vertices is called a *tree*. A tree may therefore be thought of as 'efficient' in that it has just enough edges to allow communication between each pair of vertices. In this section the concepts of connected graph, tree and binary tree are developed.

8.2.1 Connected graphs

Definition 8.14 (Connected) *A graph G is* connected *if, for every pair of vertices x, y in G, there is a path $P(x, y)$.*

Let e be an edge of a graph G. $G - \{e\}$ is the graph obtained by deleting e from the edge set of G.

Definition 8.15 (Bridge) *Let G be a connected graph. An edge e, of G, is called a* bridge *if the graph $G - \{e\}$ is not connected.*

Definition 8.16 (Euler walk, circuit) *A walk that contains each edge exactly once is called an* Euler *walk (or trail). A circuit that contains each edge exactly once is called an* Euler *circuit.*

Leonhard Euler (1707-83) was a Swiss mathematician who solved a problem about a graph in 1736. His solution was probably the first example of a 'graph theory' approach by a mathematician. Euler's problem is called the Königsberg bridge problem. It involves determining whether the general graph of Figure 8.10 has an Euler circuit.

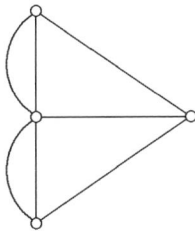

Figure 8.10: Königsberg bridge graph

Euler argued that no circuit was possible, because for such a circuit to exist each vertex must be of even degree. Each edge must be used exactly once in an Euler circuit and in traversing the circuit each edge entering a vertex must be matched by a different one that leaves the vertex; so the degree of the vertex must be a multiple of 2. The argument applies equally to all vertices.

The vertices in the Königsberg bridge graph are all odd. So the problem has no solution.

Theorem 8.3 (Euler) *If a graph has an Euler circuit then each vertex must be of even degree. That is, a necessary condition for a graph to have an Euler circuit is that each vertex be of even degree.*

Proof: Euler's argument (above) proves the theorem.

Corollary: *If a graph has an Euler walk then it must have exactly two vertices of odd degree or no vertices of odd degree.*

Proof: Let G have an Euler walk starting at vertex a and ending at vertex b.

If $a = b$, then theorem 8.3 holds and all vertices have even degree.

If $a \neq b$, make a graph $G \cup \{e\}$ by adding the edge $e = (a, b)$ to G.

The new graph has an Euler circuit consisting of the Euler walk from a to b in G together with the edge e from b to a.

By theorem 8.3 every vertex in $G \cup \{e\}$ is even.

Remove e, to obtain G. Now every vertex is even, except a, b which each have had one incident edge removed and are therefore odd.

A converse of Euler's Theorem is true, and could be proved by mathematical induction.

Theorem 8.4 (Euler converse) *If every vertex of a finite, connected graph has even degree then the graph has an Euler circuit. That is, a sufficient condition for a graph to have an Euler circuit is that every vertex should have even degree.*

Corollary: *A finite, connected graph that has exactly two vertices of odd degree must have an Euler path.*

The following algorithm to construct an Euler walk or circuit in a graph is due to Fleury. An informal, 'pencil and paper', description is:

Start at any vertex if they are all of even degree, or at a vertex of odd degree, if exactly two have odd degree. Traverse the edges in an arbitrary manner, subject only to the following rules:

1. erase the edges as they are traversed, and if any vertices become disconnected, erase them too;

2. choose a bridge as an edge only if there is no alternative.

Algorithm 8.1 (Fleury) *To construct an Euler walk or an Euler circuit for a graph.*

input:	A graph G.
	Precondition: $V(G)$ is finite and G is connected and (G has exactly two vertices of odd degree or all even vertices)
output:	A walk W.
	Postcondition: (W is an Euler walk for G and G has 2 odd vertices) or (W is an Euler circuit for G and all vertices of G are even).

method: If the vertices of G are all even call one of them a.

If two vertices are odd call one of the odd vertices a.

Let W be the graph containing vertex a (alone).

While G has edges left:

if degree$(a) > 1$ then

find an edge $\{a, x\}$ in G which is not a bridge

remove edge $\{a, x\}$ from G

add edge $\{a, x\}$ and vertex x to W

if degree$(a) = 1$ then

(there is only one edge $\{a, x\}$ incident with a)

remove the edge $\{a, x\}$ from G; remove a from G

add $\{a, x\}$, a and x to W

Let the new vertex 'a' be the last 'x'

(Now repeat. A new x is found in the first line of the loop.)

Definition 8.17 (Hamiltonian path, cycle) *A* Hamiltonian path *in a simple graph is one that contains each vertex exactly once.*
A Hamiltonian cycle *contains each vertex exactly once, except that the first vertex is the same as the last.*

A necessary and sufficient condition for an Euler circuit is that each vertex should be of even degree. No such condition is available for Hamiltonian cycles. Finding a Hamiltonian cycle in a graph is equivalent to the 'travelling salesman problem' (with edges of unit weight) discussed in section 7.1.2 and no efficient algorithm is known for its solution.

Definition 8.18 (Disconnected) *A graph that is not connected is said to be* disconnected. *Subgraphs of the graph that are connected are called* components *of the graph.*

In Chapter 3 (p. 57) an equivalence relation was defined to be a binary relation with three properties. It must be reflexive, symmetric and transitive. An *equivalence relation* on a set S partitions S into subsets called 'equivalence classes'. Each pair of elements in a given equivalence class satisfies the relation. Now we define a relation called 'connected' on vertices and use this relation to identify the connected components of a graph.

Theorem 8.5 *Let ρ be the relation defined on the vertices of a simple graph G*

$$x \rho y = \text{ 'there is a path } P(x, y)\text{'}$$

Then ρ is an equivalence relation on the vertices of G.

Proof :   **1.** ρ **is reflexive.** There is a path (of length zero) P(x,x) for all x.

Hence $x\rho x$.

2. ρ **is symmetric.** If there is a path P(x,y) then there is also a path P(y,x).

Hence, if $x\rho y$ then $y\rho x$.

3. ρ **is transitive.** Given $x\rho y$ and $y\rho z$ there is a path P(x,y) and a path Q(y,x).

Putting them together there is certainly a walk from x to z, but it may have repeated vertices and so may not be a path.

Go from x to y along P until reaching the first vertex common to P and Q. Call this vertex u. There must be a 'u' since y belongs to P and Q even if no other vertex does.

Now the portion of P from x to u together with the portion of Q from u to z gives a path from x to z.

Hence $x\rho z$.

Therefore ρ is an equivalence relation.

We deduce from the statement 'ρ is an equivalence relation on the vertices of a graph', that it partitions the vertices of G into equivalence classes. Each equivalence class of vertices 'joined by a path' forms a connected subgraph of G with those paths. The connected subgraphs are the components of G (by definition 8.18).

8.2.2   Trees

Starting from the definition below additional properties of a tree are derived, leading to theorem 8.8, which collects the properties together. The theorem provides us with six equivalent definitions for a tree. The following definition is used by many writers.

Definition 8.19 (Tree) *A tree is a connected graph with no cycles.*

Theorem 8.6 *Let a tree have p vertices and q edges. Then $p = q + 1$.*

Proof: The proof is by mathematical induction on p.

For $p = 1$ there is one tree. It has one vertex and no edges. That is $p = 1$ and $q = 0$, which satisfies the theorem.

Suppose the theorem is true for all trees with fewer than p vertices. Let T be any tree with p vertices.

Choose any edge $e = (x, y)$ of T.

The graph $T - \{e\}$ is not connected. For if it were, there would be a path $P(x, y)$ in $T - \{e\}$ which together with e would give a cycle in T. And that contradicts the definition of 'tree'.

Thus e is a bridge of T and $T - \{e\}$ has two components.

Each of the components is a tree with fewer than p vertices.

By the inductive assumption the theorem holds for each component.

Let one component have q_1 edges and the other have q_2 edges. Then:

$$p = (q_1 + 1) + (q_2 + 1) = (q_1 + q_2 + 1) + 1 = q + 1$$

recalling that T has the extra edge e.

Thus by mathematical induction the theorem is true for all integers $p \geq 1$.

Theorem 8.7 *If x and y are distinct vertices in a tree T then there is exactly one path $P(x, y)$.*

Proof: By definition T is connected.

Thus for all vertices x, y, there must be a path $P(x, y)$.

But there cannot be a second path $Q(x, y)$, for if this were so a cycle could be constructed from PQ.

The converse of the last theorem is also true. At this point we will collect the properties of trees into a composite theorem which includes the two theorems above.

Theorem 8.8 (Trees) *If any of the following statements is true for a graph T with n vertices, then all the statements are true.*

1. *T is a tree.*

2. *T contains no cycles and has $n - 1$ edges.*

3. *T is connected and has $n - 1$ edges.*

4. *T is connected and every edge is a bridge.*

5. *Every pair of vertices in T is connected by exactly one path.*

6. *T contains no cycles, but the addition of a new edge (with no new vertices) creates a cycle.*

The point of theorem 8.8 is that any of the statements **2** to **6** contained in it could replace definition 8.19 as a satisfactory definition of a tree.

The logical organisation of the proof is interesting. Each of $1 \rightarrow 2$, $2 \rightarrow 3, 3 \rightarrow 4$, $4 \rightarrow 5, 5 \rightarrow 6$, $6 \rightarrow 1$ is shown to be true. Then any statement implies all the others and Theorem 8.8 is proved.

Proof: The theorem is true for $n = 1$. Consider $n \geq 2$.

$1 \rightarrow 2$. This follows from definition 8.19 and theorem 8.6.

$2 \rightarrow 3$. Suppose T is disconnected. Obtain a contradiction from theorem 8.6.

$3 \rightarrow 4$. The removal of any edge results in a graph with n vertices and $n - 2$ edges. It may be shown, by induction on n, that such a graph is disconnected. So by contradiction the proposition is true.

$4 \rightarrow 5$. Since T is connected, every pair of vertices is connected by a path. But no pair of vertices is connected by two paths because the two paths would produce a cycle; and that contradicts the statement that each edge is a bridge.

$5 \rightarrow 6$. If T contained a cycle then at least two vertices would be connected by two paths; so T cannot contain a cycle. The addition of a new edge between vertices $u, v \in T$ creates a cycle with the existing path $P(u, v)$.

$6 \rightarrow 1$. Suppose T is disconnected. Add an edge with one vertex in one component and the other vertex in a different component. No cycle is created, and that contradicts the proposition.

8.2.3 Binary trees

Any vertex of a tree may be chosen as a root vertex. But this flexibility is removed if we demand that the graph be *directed*. A directed tree may be restricted by placing a limit on the number of edges 'leaving' each vertex. The simplest such structure is a *binary directed tree* in which no more than two edges may 'leave' each vertex. If, further, the edges leaving each vertex are *ordered* by labelling them (say) 'left' and 'right', then each vertex is located by a unique path represented by a sequence of 'left' and 'right' turns from the root.

Such a tree might be called a 'binary, directed, ordered tree' which is unfortunately rather a lot to say or write. You will see from Figure 8.11, however, that it is not too hard to visualise. Let us call it a 'BDOT' to save writing.

A *subtree* of a tree (t) is some subset of the vertices of t together with a subset of the edges of t, that form a tree. Each vertex of a binary directed ordered tree t is the root of a subtree of t and this fact suggests a recursive definition.

Definition 8.20 (BDOT) *A binary, directed, ordered tree (t) is defined recursively by*

$$t = null \vee (x, y, z)$$

where 'null' is the empty tree, having no vertices or edges and (x, y, z) is an ordered triple.

x *is the 'left subtree';*
y *is the root of t; and*
z *is the 'right subtree'.*

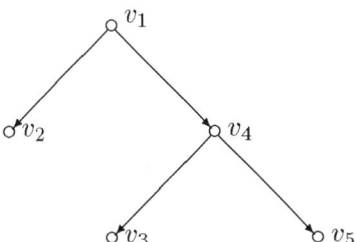

Figure 8.11: A binary, directed, ordered tree

The second element (y) of the ordered triple, called the root of the tree, may be used to store 'information' and one possible task is to search a tree for the information stored in it.

Example 1: Show that the graph pictured in Figure 8.11 satisfies definition 8.20 for a binary, directed, ordered tree.

Solution: The graph pictured in Figure 8.11 is a binary directed ordered tree if it is *null* or $(v_2, v_1, v_3 v_4 v_5)$ is a tree.

The graph is not *null* so we must investigate the ordered triple. The triple is a 'BDOT' if v_2 is a 'BDOT' (the left subtree) and v_1 is the root (it is, since it is a vertex in second place in the triple), and if (v_3, v_4, v_5) is a 'BDOT' (the right subtree).

v_2 is a BDOT if $(null, v_2, null)$ is a 'BDOT', which it is.

(v_3, v_4, v_5) is a 'BDOT' if:

1. v_3 is a 'BDOT' (it is because $(null, v_3, null)$ is a 'BDOT'); and
2. v_4 is the root (it is); and
3. v_5 is a 'BDOT' (it is because $(null, v_5, null)$ is a 'BDOT').

Therefore the graph pictured in Figure 8.11 is a binary, directed ordered tree.

To help in the extraction of information we define the following partial functions on a tree (t). The functions are not defined for the *null* tree. Otherwise:

$$
\begin{aligned}
left(t) &= x, \quad \text{the left subtree of } t \\
info(t) &= \quad \text{the information at } y, \text{ the root of } t \\
right(t) &= z, \quad \text{the right subtree of } t
\end{aligned}
$$

There are several ways in which information stored at the vertices of a binary directed ordered tree may be extracted. The following recursively defined function inputs a binary directed ordered tree (t), with information held at each vertex and outputs the information as a list.

Call the function '*Tlist*'; call the set of binary, directed, ordered information trees 'T' and call the set of lists of information 'L.' Then:

$$Tlist: \quad T \longrightarrow L$$

$$
Tlist(t) = \begin{cases}
<> & \text{if } t = null \\
\left. \begin{array}{l} Tlist(left(t)) \\ \text{append } < info(t) > \\ \text{append } Tlist(right(t)) \end{array} \right\} & \text{if } t \neq null
\end{cases}
$$

Example 2: Apply *Tlist* to the binary directed ordered information tree pictured in Figure 8.11. The information stored in the tree is 'r', 'p', 'i', 'z', 'e' located at the vertices v_1, v_2, v_3, v_4, v_5, respectively.

The output is:

$Tlist(v_2) = <> \; || < p > \; || <>$

append $< r >$

append $Tlist(v_3, v_4, v_5)$

   (i.e. $<> \; || < i > \; || <> \; || < z > \; || <> \; || < e > \; || <>$)

Carrying out the appends we obtain $< p, r, i, z, e >$.

The function *Tlist* may be written as an algorithm whose purpose is the extraction of information stored at the vertices of a binary directed ordered tree.

Algorithm 8.2 (Binary information tree) *To obtain a list of the information stored in a binary directed ordered tree.*

input:	t, a tree.
	Precondition: t satisfies definition 8.20
output:	l, a list.
	Postcondition: the list l contains the information from each vertex of t in some order.
method:	*Tlist(t)*.

8.3 SPANNING TREES

In this section we look at a particular kind of tree called a *spanning* tree. Spanning trees are important because they have the least number of edges while remaining connected.

Definition 8.21 (Spanning tree) *A subgraph of a connected graph G is called a* spanning *tree if and only if it is a tree that has the same vertex set as G.*

Algorithm 8.3 (Spanning tree) *To construct a spanning tree of a finite simple graph G, if one exists.*

input:	G a simple graph with n vertices
output:	T, a tree.
	Postcondition: $V(T) = V(G)$ or 'G is not connected'.
	That is, T has the same vertices as the input graph G and is a spanning tree if G is connected.
method:	Choose any vertex u_0 in G to be the first vertex of T.
	While there is an edge of G incident with a vertex (u) in T and a vertex (v) not in T:
	add this edge (u, v) to T,
	add v to T.
	If T contains n vertices then T is a spanning tree for G.
	If T contains fewer than n vertices then G is not connected.

Proof: Initially T has one vertex. For each pass through the 'while loop', one edge and one vertex are added to T. Therefore at the end of a pass through the 'while loop' the number of vertices is one more than the number of edges.

Also T is connected.

Therefore by theorem 8.8, T is a tree after each pass through the 'while loop'; including the last.

If T has n vertices, then it has all the vertices of G and is a spanning tree of G.

If T has fewer than n vertices when the algorithm terminates then there must be at least one vertex in G but not in T. Let s be such a vertex.

Suppose G is connected. (We will see that this leads to a contradiction.) Then there is a path $P(u_0, s)$ in G. Let the vertices of this path be $s_0, s_1 \cdots s_k$ where $s_0 = u_0$ and $s_k = s$.

Now $k \geq 1$ since $u_0 \neq s$. Let s_i be the first vertex not in T. (There must be such a vertex since s_k will do if there is no other.) And $i \geq 1$ since $u_0 \in T$. Now (s_{i-1}, s_i) is an edge that joins a vertex in T to a vertex not in T.

But the algorithm terminated because no such edge exists, so G must be *not* connected.

Time complexity The algorithm at worst passes through the 'while loop' once for each edge in T, that is $n - 1$ times. The algorithm has time complexity $n - 1$ and is $O(n)$.

8.3.1 Search trees

Spanning trees have the useful property that there is a unique path between each pair of vertices. However, although algorithm 8.3 produces a spanning tree efficiently, there is no way to find a path between an arbitrary pair of vertices without a time-consuming search.

The difficulty can be avoided in the following way. Call the initial vertex u_0 in algorithm 8.3, the *root* vertex. Consider the step in the algorithm that adds the edge (u, v) to the tree T where $u \in T$ and $v \notin T$. The vertex u may be spoken of as the 'predecessor' or 'parent' of the vertex v in T, and we write

$$pr(v) = u.$$

Each vertex except the root vertex has a parent in T. For completeness define an 'empty vertex', $\emptyset$. Then the 'parent' of the root vertex is $\emptyset$.

The function pr is now defined on all the vertices of T. The function is represented on a diagram by 'pointers' or arrows as in Figure 8.12. In that diagram:

$$pr(a) = \emptyset, pr(b) = a, pr(c) = a, pr(d) = b, pr(e) = b, pr(f) = b$$

and the root vertex is a.

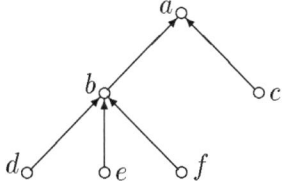

Figure 8.12: The 'parent' function

The parent or pointer function may be applied successively to vertices to reach the root vertex, for example in Figure 8.12:

$$pr^2(f) = pr \circ pr(f) = pr(b) = a, \text{ the root vertex.}$$

Definition 8.22 (Level or depth) *If $pr^k(v) = r$ where r is the root vertex then the vertex v is said to be at level (or depth) k. Write $level(v) = k$ and define $level(r)$ to be 0.*

A path between two vertices m, n may now be found by searching through the 'parents', 'grandparents', etc. of m, n until a common ancestor is found.

A spanning tree with the additional structure given by the 'parent' function is called a *search tree*.

In the following algorithm vertices that are adjacent only to vertices in the tree are said to be *done*. The algorithm terminates when all vertices are done. The algorithm makes use of a structure called a *queue*. A queue is a list of objects waiting to be processed in some way. The rule by which the objects are chosen for removal and processing is called the 'priority' of the queue and in the following algorithm the priority is 'first in, first out'. That is, each arrival is 'appended' to the queue and the next object for processing is the 'head' of the queue. (See Chapter 6 for definitions of 'head' and 'append'.)

Algorithm 8.4 (Breadth first search tree) *To construct a* breadth first search tree *of a finite connected simple graph.*

input: A simple graph G.

Precondition: G is connected.

output: A search tree, T.

Postcondition: T is a tree and $V(T) = V(G)$; that is T has the same vertices as the input graph G and is a spanning tree for G.

method: Let the tree T be the empty tree. ($T \leftarrow null$)

Choose a root vertex u_0, and add it to T.

$pr(u_0) \leftarrow \emptyset$ ('empty vertex')

Let the queue of vertices be $< u_0 >$.

While the queue is not empty or $V(T) < V(G)$:

    call the vertex at the head of the queue, 'v'

    for each vertex w that is adjacent to v in G and is not in T

        add (v, w) to T

        add w to T

        append w to the queue of vertices

        $pr(w) \leftarrow v$

    remove v from the queue (it is said to be 'done')

Proof: Algorithm 8.4 is only a refinement of algorithm 8.3 in which vertices are chosen in a particular way and a pointer function is added. Therefore if G is connected, algorithm 8.4 finds a search tree with root u_0 for G.

The algorithm is $O(n)$ in time where n is the number of vertices of G.

Example: Construct a breadth first search tree of the graph G given the following adjacency list table.

0	1	2	3	4
1	0	0	1	1
2	2	1		
	3			
	4			

Solution: Choose '0' to be the root vertex. Then the following table shows the growth of the tree.

T	queue	parent function
0 (root)	$< 0 >$	$pr(0) = \emptyset$
(0,1), 1	$< 0, 1 >$	$pr(1) = 0$
(0,2), 2	$< 0, 1, 2 >$	$pr(2) = 0$
0 is done	$< 1, 2 >$	
(1,3), 3	$< 1, 2, 3 >$	$pr(3) = 1$

(1,4), 4	$< 1,2,3,4 >$	$pr(4) = 1$
1 is done	$< 2,3,4 >$	
2 is done	$< 3,4 >$	
3 is done	$< 4 >$	
4 is done	$<>$	

The diagram of Figure 8.13 shows the breadth first search tree of G constructed above. The 'depth' or 'level' of any vertex in the search tree may be recovered using the parent (i.e pointer) function. For example, the level of vertex 3 is obtained by observing $pr(3) = 1; pr(1) = 0$ (the root) and therefore $pr^2(3) = 0$ and vertex 3 is at level 2.

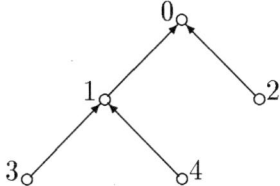

Figure 8.13: Breadth first search tree of G

Another approach to constructing a search tree is to construct paths as long as possible by searching deeper levels of the graph from the root vertex. When it becomes impossible to extend the tree further 'downward' from the root, we 'backtrack' one level to the previous vertex and try to go down a new path. The algorithm terminates when every vertex has been visited. The following algorithm to construct a *depth first* search tree uses a queue with a different priority to that of algorithm 8.4. The priority of the queue is 'last in, first out'. Each arrival is placed at the 'head' of the queue and each departure is the current 'head' of the queue.

Algorithm 8.5 (Depth first search tree) *To construct a* depth first search tree *of a finite connected simple graph.*

    input: A simple graph G.
                 Precondition: G is connected.
    output: A search tree, T.
                 Postcondition: T is a tree and $V(T) = V(G)$;
                 that is T has the same vertices as the input graph G and
                 is a spanning tree for G.

method: Let the tree T be the empty tree. ($T \leftarrow null$)
Choose a root vertex u_0, and add it to T.
$pr(u_0) \leftarrow \emptyset$ ('empty vertex')
Let the queue of vertices be $< u_0 >$.
While the queue is not empty or $V(T) < V(G)$:
  call the vertex at the head of the queue, 'v'
  if vertex w is adjacent to v in G and is not in T
    then add (v, w) to T
      add w to T
      place w at the head of the queue
      $pr(w) \leftarrow v$
    else remove v from the head of the queue

Proof: Algorithm 8.5 is also a refinement of algorithm 8.3 in which a
pointer function is added. Therefore if G is connected, algorithm 8.5 finds a search tree with root u_0 for G.
The algorithm is $O(n)$ in time where n is the number of vertices of G.

Example: Use algorithm 8.5 to find a depth first spanning tree for the
graph G whose adjacency list table is given below.

0	1	2	3	4
1	0	1	1	0
4	2	3	2	
	3			

Solution: The application of algorithm 8.5 to G, above, is laid out
below.

T	queue	parent
0	$< 0 >$	$pr(0) = \emptyset$
(0,1), 1	$< 1, 0 >$	$pr(1) = 0$
(1,2), 2	$< 2, 1, 0 >$	$pr(2) = 1$
(2,3), 3	$< 3, 2, 1, 0 >$	$pr(3) = 2$
	$< 2, 1, 0 >$	
	$< 1, 0 >$	
	$< 0 >$	
(0,4), 4	$< 4, 0 >$	$pr(4) = 0$

All the vertices of G are now in T, so stop. The search tree is pictured
in Figure 8.14.

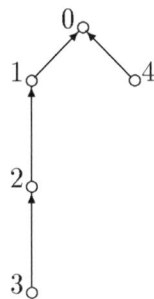

Figure 8.14: Depth first search tree

8.3.2 Minimum weight spanning tree

Given a natural gas field with wells and a treatment/pumping plant it is required to lay pipes so that each well is connected to the plant at least total cost. Since each well is connected to the plant the system is connected. And for economy there should be no cycles. The problem may be modelled as a graph by taking the wells and the plant as vertices and the possible pipe connections as edges, each labelled with its construction cost. Then the problem is to find a spanning tree for the graph with minimum total cost.

Problem 8.1 (Minimal spanning tree) *How can we find a spanning tree for an edge-labelled graph to minimise the sum of the labels (weights) on the edges of the tree?*

Algorithm 8.6 (Kruskal) *To construct a spanning tree of least total edge-weight, for a connected graph G.*

    input: A graph G with edges $e_1, e_2, \cdots e_m$
               and n vertices and edge weights $w(e_1), \cdots w(e_m)$.
               Precondition: $V(G)$ is finite and G is connected.
    output: A tree, T.
               Postcondition: T is a tree with the minimum sum of edge weights.
               $V(T) = V(G)$.

method: Order the edges of G so that

$$w(e_1) \leq w(e_2) \leq \cdots \leq w(e_m).$$

Set T to be e_1 together with its vertices.

While the number of edges in T is less than $n - 1$:
  if T with the next edge in order has no circuits,
    then add the next edge and its vertices to T

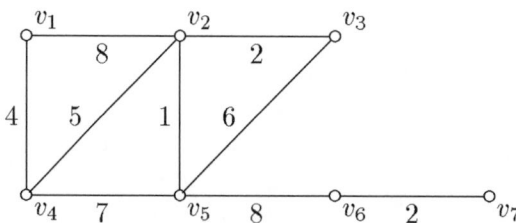

Figure 8.15: An edge-weighted graph

Example: Find the minimum weight spanning tree for the graph in Figure 8.15.

Solution: Comparing the weights for the edges of g:

$$1 \leq 2 = w(v_2, v_3) \leq 2 = w(v_6, v_7) \leq 4 \leq 5 \leq 6 \leq 7$$

$$\leq 8 = w(v_1, v_2) \leq 8 = w(v_5, v_6)$$

Where edges have equal weights the choice of first edge is arbitrary. The spanning tree contains the edges $(v_2, v_5), (v_2, v_3), (v_6, v_7), (v_1, v_4),$

(v_2, v_4), not (v_3, v_5), not (v_4, v_5), not (v_1, v_2), (v_5, v_6). The algorithm stops because six edges have been placed in the tree.

In the above application of algorithm 8.5, circuits have been identified visually from the picture. The algorithm would need to be refined to allow automatic identification of circuits, before use on a large problem.

8.4 EERCISES

1. The vertices adjacent to each of the vertices $1, 2, 3, 4, 5, 6$ in a graph G, are listed in the following table:

1	2	3	4	5	6
3	4	1	2	1	2
5	6	5	5	3	3
		6		4	

(a) Is it possible to find an Euler walk for G?

(b) Construct a breadth first search tree for G from the vertex labelled 1.

(c) Find the vertex or vertices furthest from the vertex labelled 1.

(d) Construct a tree by depth first search from vertex 1.

2. The vertices adjacent to each of the vertices $1, 2, 3, 4, 5, 6$ in a graph G, are listed in the following table:

1	2	3	4	5	6
2	1	5	1	2	1
4	5			3	2
6	6			6	5

(a) Is it possible to find an Euler walk for G? If so, write it down.

(b) Is it possible to find a Hamiltonian path for G? If so, write it down.

(c) Construct a breadth first search tree from the vertex v_4. Find the vertex or vertices furthest from the vertex v_4

(d) Find a depth first search tree from v_4.

3. The adjacency matrix for a graph G with vertices $v_1, \ldots, v_6$ is:

$$A = \begin{pmatrix} 0 & 0 & 1 & 0 & 1 & 1 \\ 0 & 0 & 0 & 1 & 0 & 1 \\ 1 & 0 & 0 & 1 & 1 & 1 \\ 0 & 1 & 1 & 0 & 0 & 1 \\ 1 & 0 & 1 & 0 & 0 & 0 \\ 1 & 1 & 1 & 1 & 0 & 0 \end{pmatrix}$$

(a) Find an Euler walk for G.

(b) Construct a tree of shortest paths from v_4.

(c) Find the vertex, or vertices, furthest from v_4.

4. Three pirates and three sailors are on the left bank of a river and wish to cross to the right bank where a treasure is buried. There is one boat which will carry two people. The boat can be rowed by a single person. The river is infested with alligators. If the pirates outnumber the sailors they will throw them in the river, otherwise the sailors are safe. Model the problem with a graph and devise a sequence of moves to transfer the six people safely across the river.

5. Draw graphs corresponding to the adjacency matrices:

$$A = \begin{pmatrix} 0 & 1 & 1 & 0 \\ 1 & 0 & 0 & 1 \\ 1 & 0 & 0 & 1 \\ 0 & 1 & 1 & 0 \end{pmatrix}, \quad B = \begin{pmatrix} 0 & 1 & 0 & 1 & 0 \\ 1 & 0 & 1 & 0 & 1 \\ 0 & 1 & 0 & 1 & 0 \\ 1 & 0 & 1 & 0 & 1 \\ 0 & 1 & 0 & 1 & 0 \end{pmatrix}$$

Calculate A^2 and B^2 and interpret the information they give about walks in the graphs.

6. Computers in six locations on campus are to be linked by coaxial cable. Locations that are not directly linked may communicate via one or more intermediate computers. Find the minimum cost of linking the computers.

		B	C	D	E	F
Cost in $ '000	A	10	20	25	25	12
of links	B		10	12	14	10
	C			5	8	12
	D				11	18
	E					6

7. The vertices of a graph are labelled with four-bit binary strings

$$(0000), (0001), \cdots (1111).$$

Two vertices are joined by an edge if and only if their labels differ by exactly one bit. Thus (0101) is adjacent to (1101) but not to (0110). Carry out a 'breadth first' search to construct a spanning tree from root vertex (0101).

The following questions are suitable for investigation and for a short tutorial presentation.

8. Let G be a *bipartite* graph. A bipartite graph is one whose vertices are partitioned into two sets A, B and each edge of G contains one vertex from A and the other from B.

Prove that, if M is the adjacency matrix of G then M^i, where i is odd, has zeros on all the diagonal entries. (A proof in words based on Theorem 8.2 is perhaps the best approach.)

9. Show that the graph whose adjacency matrix is

$$A = \begin{pmatrix} 0 & 1 & 0 & 0 & 0 \\ 1 & 0 & 1 & 0 & 0 \\ 0 & 1 & 0 & 0 & 0 \\ 0 & 0 & 0 & 0 & 1 \\ 0 & 0 & 0 & 1 & 0 \end{pmatrix}$$

is disconnected.

Ask your tutor or lecturer about *partitioned* matrices. Then try to prove by mathematical induction that A^k has zero entries for all k. You should be able to show that v_1, v_2, v_3 and v_4, v_5 are two components of the graph that are not connected.

10. Define a 'ternary, directed, ordered tree', along the lines of definition 8.20. Modify the function *Tlist* defined on page 174 to fit your definition.

11. Describe the effect on algorithm 8.2 produced by changing the order of the statements in the function *Tlist* (p. 174).

12. Kruskal's algorithm 8.5 includes the predicate 'T with the next edge has no circuits'. Design an algorithm to test for the presence of a circuit.

13. My *Quantitative Business Package* used in operations research problems has a program to construct a minimum edge-weight spanning tree for a graph G. The package splits the graph into two components. The first component is a tree of minimum edge-weight for a subgraph of G . The second component C consists of vertices and edges of G which have not yet been considered for placement in the tree T. The algorithm finds the edge with least edge-weight from the set of edges of G that have one vertex in T and one vertex in C, and adds this edge and its C-vertex to T. Design an algorithm for the package.

14. Aho, Hopcroft and Ullman (1984) state and prove (their theorem 5.2) that algorithm 8.4 to construct a depth first spanning tree requires $O(\text{Max}(n, e))$ steps on a graph with n vertices and e edges. What does this mean? Give the proof.

15. Find the time complexity for algorithm 8.3 to construct a breadth first spanning tree.

Chapter 9

Counting

Calculation of the time-complexity of an algorithm requires counting the number of times some basic step or operation is repeated. The determination of a theoretical probability usually requires counting the number of elements in a subset. Counting is also important for less direct reasons. For example, counting steps in an algorithm gives insight into the structure of the algorithm. In counting we seek to identify, or impose, some structure on the set of objects being counted. This chapter is about the process of counting and learning to use techniques that simplify counting. The chapter concludes with algorithms for the generation of 'arrangements' of a set of objects and for the generation of subsets of a set.

Most people think of counting as their first mathematical experience. The ability to count is regarded as a 'basic skill'. Nevertheless counting is not easy to define satisfactorily. Reciting the names 'one, two, three, $\cdots$' is not of itself counting, nor even necessary to counting, because a person using a different language would use different names.

At a deeper level, the discovery of the reflexive paradox (that the set of all sets which are not members of themselves is self-contradictory) led to new approaches to the foundations of mathematics. The constructive approach begins with 'recursive' arithmetic and it is this approach that seems to offer the most to computing.

9.1 TECHNIQUES OF COUNTING

9.1.1 Numbers and counting

A natural number x in this constructive scheme, is defined by:

x is a number if, and only if, it is '0' or 'the successor of a number'.

We are asked to accept that '0' is a number, and if x is a number then '*successor*(x)'$= x + 1$ is also a number. The whole numbers are then defined constructively, that is by an algorithm, to be the sequence:

   0, *successor*(0), *successor*$(\text{successor}(0))$, $\cdots$

The *names* of the numbers depend on the translation scheme. For example we may translate the above sequence to:

$$0, 1, 2, \cdots$$

$$\text{or } 0, 1, 10, \cdots \text{ (binary)} \quad \text{or} \quad \text{null, ein, zwei } \cdots$$

We have been asked to accept very little. Yet with the inclusion of the arithmetic operations, also defined recursively, all the processes of classical arithmetic may be developed. For example see Goodstein (1957).

For counting to begin it is necessary that the objects to be included in the count be identified unequivocally (possibly by identifying the objects to be excluded from the count). Therefore the objects to be counted constitute a set.

Techniques of counting have been developed to exploit particular structures imposed on the set and so may be classified by that structure. It should be noted that the definition of number produces natural numbers and the process of counting may only be applied to *finite* sets.

9.1.2 Counting by tally

The tally method of counting assumes nothing about the structure of the set to be counted.

Let A be the set to be counted. The number of elements in A is written $n(A)$ or $|A|$. Counting the number of elements in a set A is defined recursively by:

$$n(A) = \begin{cases} 0 & \text{if } A = \emptyset \\ n(A - \{l\}) + 1 & \text{if } A \neq \emptyset \end{cases}$$

where l is an element of A.

The statement '$n(A - \{l\}) + 1$' is just the *successor* function, which from the definition of number in the section above is defined for all natural numbers and 0. The statement $n(\emptyset) = 0$ declares that the number of elements in the empty set is zero, a very reasonable assertion.

The idea behind the definition of counting is that of 'crossing off' each item as it is 'counted' or 'tallied'. Defining 'counting' as a function (from sets to numbers) is equivalent to providing an algorithm for counting the elements of a set. The counting algorithm is illustrated informally in the examples below.

Example 1: Count the number in the set $A = \{a, b, c\}$.

$$
\begin{aligned}
n(A) &= n(\{a, b\}) + 1 \\
&= (n(\{a\}) + 1) + 1 \\
&= ((n(\emptyset) + 1) + 1) + 1 \\
&= ((0 + 1) + 1) + 1
\end{aligned}
$$

The *process* of counting is complete. To represent the number as a numeral, a translation routine is needed. In English this may give 'three', in French 'trois' or in a computing laboratory 11 (binary).

$$
\begin{aligned}
&counter \leftarrow 0 \\
&\text{while } condition \\
&\qquad \text{action 1} \\
&\qquad \text{action 2} \\
&\qquad counter \leftarrow counter + 1
\end{aligned}
$$

Figure 9.1: Algorithm fragment

Example 2: In the algorithm of Figure 9.1 the variable *counter* counts the number of times that the algorithm passes through the 'while' loop.

The number produced by *counter* when the algorithm is executed, reflects faithfully the definition of number given above.

9.1.3 Counting members of a sequence

Suppose the elements of the set A to be counted form a sequence. Then it should be possible to find a one-one function that maps the set A onto the set $\{1, 2, 3, \cdots, n\}$ and the number of elements in A will be n.

Example 1: Let m, n be integers such that $m \leq n$. How many integers are there from m to n inclusive?

Solution: The question asks us to count the members of the sequence $m, m + 1, m + 2, \cdots, n$. The function 'subtract m' followed by the function 'add 1' maps the sequence one-one onto the set

$$
\{1, 2, 3, \cdots, (n - m + 1)\}.
$$

Therefore the number of integers in the original sequence is $n - m + 1$.

Example 2: How many two-digit numbers are multiples of 4?

   Solution: The set of two-digit numbers may be arranged as the sequence

$$S \; = \; < 10, 11, 12, \cdots 99 >$$

The function 'divide by four' maps the sequence one-one onto

$$S' \; = \; < 2.5, 2.75, \cdots, 24.75 >$$

The multiples of four in the sequence S correspond one-one to the integers in S'. Define 'the set of integers in S' to be S'' where

$$S'' \; = \; < 3, 4, \cdots, 24 >$$

The set S'' is mapped by the function 'subtract 2' onto the the set

$$\{1, 2, \cdots, 22\}$$

Therefore there are 22 two-digit multiples of four.

Example 3: Count the number of elements in the set

$$A = \{93, 95, \cdots, 855\}$$

consisting of all the odd numbers from 93 to 855 inclusive.

Solution: The function 'subtract 91' maps A one-one onto

$$A' = \{2, 4, 6, \cdots 764\}$$

The function 'divide by 2' maps A' one-one onto

$$A'' = \{1, 2, 3, \cdots, 382\}$$

Therefore there are 382 numbers in A.

9.1.4   Counting Cartesian products

Theorem 9.1 *The number of elements in the Cartesian product $A \times B$ is $n(A) \times n(B)$.*

This theorem is often referred to as the 'rule of product'. It may be extended to any finite number of sets using a proof by mathematical induction.

   The proof of theorem 9.1 itself may be based on the introduction of a variable, *counter*, in the algorithm for constructing the ordered pairs (a, b) of $A \times B$.

$$counter \leftarrow 0$$
$$\text{for } i = 1 \text{ to } n(A)$$
$$\quad \text{for } j = 1 \text{ to } n(B)$$
$$\quad\quad \text{write } (a_i, b_j)$$
$$\quad\quad counter \leftarrow counter + 1$$

For each of the $n(A)$ choices for a_i the pair $(a_i, \quad)$ is completed by $n(B)$ choices for b_j to give a pair (a_i, b_j). By the end $n(A) \times n(B)$ pairs have been constructed.

Example: How many distinct 8-digit binary sequences are possible?

Solution: Each digit in the sequence is chosen from the set $B = \{0, 1\}$ and $n(B) = 2$.

The distinct sequences correspond to the ordered 8-tuples of the Cartesian product

$$B \times B \times B \times B \times B \times B \times B \times B$$

and by theorem 9.1 (extended) there are 2^8 of them. Therefore 256 binary sequences of eight digits are possible.

9.1.5 Counting subsets

The counting techniques described in this section depend on some subset structure being given for the set to be counted. The counting techniques are applications of the following theorems.

Theorem 9.2 *Let the set A be partitioned into cells $A_1, A_2, \cdots A_s$. Then:*

$$n(A) = n(A_1) + n(A_2) + \cdots + n(A_s)$$

The theorem is sometimes called the 'rule of addition'. It may be proved by mathematical induction on s.

Theorem 9.3 *Let the set A have complement A' relative to a universal set U. Then:*

$$n(A) = n(U) - n(A')$$

The theorem is sometimes called the 'rule of complement'. It is a corollary of theorem 9.2.

Theorem 9.4 (Inclusion/Exclusion) *Let A, B, C be finite sets. Then:*

$$n(A \cup B) = n(A) + n(B) - n(A \cap B)$$

and

$$n(A \cup B \cup C) = n(A) + n(B) + n(C) - n(B \cap C) - n(C \cap A)$$
$$-n(A \cap B) + n(A \cap B \cap C)$$

The theorem called 'inclusion/exclusion' may be generalised to any number of sets. The proof is by mathematical induction on the number of subsets.

Example: How many integers from 1 to 100 have a '7' in their decimal representation?

Let A be the set of integers from 1 to 100 that have a '7' in their decimal representation.

Solution 1: Partition A into cells A_i where:

A_i is the set of numbers from $10i$ to $10i + 9$ that contain a digit 7.

Therefore $A_0 = \{7\}, A_1 = \{17\}, \cdots A_7 = \{70, 71, \cdots 79\}, \cdots$
$$A_9 = \{97\}.$$

and $n(A) = 1 + 1 + 1 + 1 + 1 + 1 + 1 + 10 + 1 + 1 = 19$

Solution 1 imposes a partition on the set and utilises theorem 9.2.

Solution 2:   Choose the universal set to be $U = \{1, 2, \cdots, 100\}$. Now U is in one-one correspondence with $\{00, 01, 02, \cdots 99\}$, the set of all pairs of digits. The correspondence is obvious, except perhaps the correspondence between 00 and 100.

$$
\begin{aligned}
n(U) &= 10 \times 10 = 100 \\
n(A') &= 9 \times 9 = 81 \\
n(A) &= n(U) - n(A') = 100 - 81 = 19
\end{aligned}
$$

Solution 2 counts the complementary set and then uses theorem 9.3.

Solution 3: Let X be the subset of A in which the 'tens' digit is 7. Let Y be the subset of A in which the 'units' digit is 7.
Then $X \cap Y$ is the set in which both 'tens' and 'units' digits are 7. Now:

$$
\begin{aligned}
n(X \cup Y) &= n(X) + n(Y) - n(X \cap Y) \\
&= 10 + 10 - 1 \\
&= 19
\end{aligned}
$$

Solution 3 imposes subsets on A that lead to the use of theorem 9.4.

9.1.6   The pigeonhole principle

The pigeonhole principle is here presented in two versions.

Theorem 9.5 (Pigeonhole principle) *If more than n pigeons are placed in n boxes then at least one box contains two or more pigeons.*

The proof is by contradiction.

Example 1: Given four integers, show that at least two of them have the same remainder when divided by 3.

    Solution: The possible remainders on dividing by three are 0, 1 and 2. Label three boxes '0', '1' and '2'. Place the four integers into the three boxes. Then one of the boxes must have at least two integers; that is at least two integers have the same remainder when divided by three.

Example 2: Given three integers, show that there must be two whose sum is even.

    Solution: Each integer is either even or odd. Apply the pigeonhole theorem with $n = 2$, the boxes labelled 'even' and 'odd' and three 'pigeons' corresponding to three integers. Then at least two integers are even or at least two integers are odd. But the sum of two even integers is even and the sum of two odd integers is even. In each case there are two integers whose sum is even.

A more general version of the pigeonhole principle extends the idea of counting by one-one mapping used in section 9.1.3.

Theorem 9.6 (Pigeonhole principle 2) *Let f be a function with domain S and range T and let S, T be finite sets. If for some integer k:*

$$\frac{n(S)}{n(T)} > k$$

then for some $t \in T$ more than k elements of S are mapped to t by the function f.

Again the proof is by contradiction.

Example 1: Let $f : S \rightarrow T$ be a function from the set S to the set T. Let $n(S) = 7$ and $n(T) = 3$. Refer to Figure 9.2. Then one of t_1, t_2, t_3 is the image of at least three members of the set $\{s_1, s_2, \cdots s_7\}$.

    For suppose none of t_1, t_2, t_3 was the image of more than two members of S. Then S could have no more than $2 \times 3 = 6$ elements.

    Contradiction: The pigeonhole principle 2 is correct for this example.

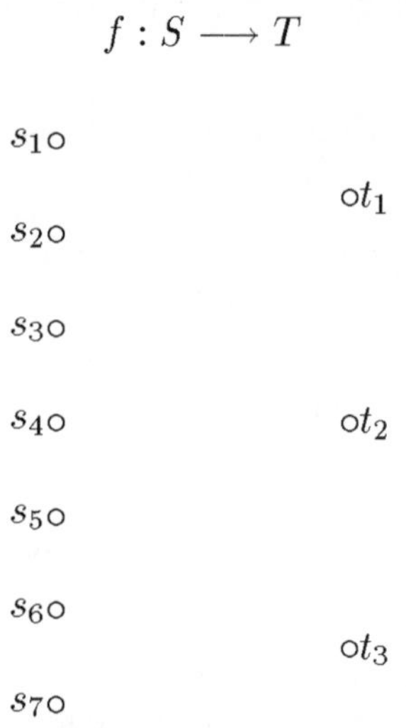

$$f : S \longrightarrow T$$

$s_1 \circ$

$\circ t_1$

$s_2 \circ$

$s_3 \circ$

$s_4 \circ$ $\circ t_2$

$s_5 \circ$

$s_6 \circ$
                               $\circ t_3$
$s_7 \circ$

Figure 9.2: Pigeonhole example

Example 2:   The numbers $1, 2, \cdots 10$ are placed around a circle in random order. Then the sum of at least one set of three consecutive numbers is at least 17.

Solution: Let the sequence of numbers around the circle be

$$a_1, a_2, \cdots, a_{10}$$

and let the sums of sets of three consecutive numbers be

$$t_1, t_2, \cdots, t_{10}.$$

The sum of the first ten numbers is 55. The following table shows sums across and down.

				Sum
a_1	a_2	$\cdots$	a_{10}	55
a_2	a_3	$\cdots$	a_1	55
a_3	a_4	$\cdots$	a_2	55
t_1	t_2	$\cdots$	t_{10}	165

Now construct a function from a set S of 165 objects to the ten 'sums of three numbers' $t_1, \cdots t_{10}$, forming a set T.

$$\frac{n(S)}{n(T)} = \frac{165}{10} > 16$$

Therefore by theorem 9.6, the pigeonhole principle, there is at least one total of three successive numbers, t_i, that is at least 17.

9.2 COMBINATORIAL REASONING

The art of combinatorial reasoning is to find a correspondence between a counting problem that we are trying to solve and a counting problem to which we know the answer. In many cases the structure we seek may be found by restating the problem using new words. Indeed combinatorial argument is often presented as a word argument rather than an algebraic argument.

9.2.1 A model for counting

The model described below appears in many forms in different questions. The model is used to identify four basic types of problem. A counting formula for each type of problem is developed.

The model Suppose we have a set of n objects $\{a_1, \cdots a_n\}$ and we wish to count the number of ways in which r objects may be chosen from the set. To specify a counting problem unambiguously it is necessary to state whether:

   1. the order of the objects chosen is important; and

   2. the same object may be chosen repeatedly;

which leads to four types of counting problem.

To simplify the explanation let us consider the choice of two objects (that is $r = 2$) from the set $\{a, b, c\}$ (that is $n = 3$).

Problem 9.1 *Order is important in the choice; repetitions are allowed.*

The choices are called *2-sequences*. The possible 2-sequences using the letters a, b, c are: $aa, ab, ac, ba, bb, bc, ca, cb, cc.$ There are three choices for each term, giving 3×3 sequences in all.

Generalisation: The number of r-sequences (order important, repetitions allowed) that may be made from a set of n distinct objects is the result of n choices for each of r positions in the sequence. There are n^r sequences in all.

Problem 9.2 *Order is important in the choice; repetitions are not allowed.*

The choices are called *2-permutations*. The possible 2-permutations that may be made from the letters of the set $\{a, b, c\}$ are $ab, ac, ba, bc, ca, cb.$ There are three ways to choose the first letter in the permutation and, for each of these, two ways to choose the second letter. There are 3×2 permutations or arrangements in all.

Generalisation: The number of r-permutations (order important, no repetition) that may be made from a set of n objects is

$$P(n,r) = n(n-1)\cdots(n-r+1)$$

Other symbols are used, for example $^{n}P_{r}$.

Example: How many ways can a club of ten people elect a president, a secretary and a treasurer?

We assume that one person cannot fill two posts; and that each permutation of three people chosen from the ten corresponds to a choice of president, secretary and treasurer.

$$P(10,3) = 10 \times 9 \times 8$$

That is, having chosen a president (in 10 ways) there are nine choices for secretary and then eight choices for treasurer.

Problem 9.3 *Order is not important; repetitions are not allowed.*

The choices are called *2-subsets*. The possible 2-subsets of $\{a,b,c\}$ are ab, ac, bc. The symbol for the number of 2-subsets that may be chosen from a set of three objects is $C(3,2)$. The letter 'C' comes from the word 'combination' which is an older name for this type of choice.

Generalisation: The number of r-subsets, often called r-combinations that may be made from a set of n objects is written $C(n,r)$. Other symbols are used, for example, $^{n}C_{r}$.

The function value $C(n,r)$ may be obtained from $P(n,r)$ by observing that r objects may be arranged in $r(r-1)(r-2)\cdots3 \times 2 \times 1$ ways. But these arrangements do not count as different in the calculation of $C(n,r)$.

$$\text{Therefore,} \quad C(n,r) = \frac{n(n-1)\cdots(n-r+1)}{r(r-1)\cdots3 \times 2 \times 1}$$

Problem 9.4 *Order is not important; repetitions are allowed.*

The choices are called *2-multisets*. The possible 2-multisets of $\{a,b,c\}$ are aa, ab, ac, bb, bc, cc. Consider three boxes labelled a, b, c (Fig. 9.3).

We have two letters to choose, but may choose two letters from the same box. Model the choice by placing two balls in the boxes. For example, two balls in box b would generate bb. A ball in a and a ball in b would generate ab. The selection bb may be read as a 'word' $SXXS$, where S represents a wall between two boxes and X represents a ball. In this notation the selection ab would be represented by the word $XSXS$.

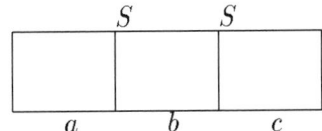

Figure 9.3: Multisets

Finding the number of selections is equivalent to finding the number of 'words' that can be made from the four symbols $SSXX$, the letters S representing the (3-1) walls separating the boxes a, b, c, and the X's representing the two balls. A word is determined by the placement of the X's in two of the $(3 - 1 + 2)$ positions. The other positions are filled automatically by S's. This can be done in $C(3 - 1 + 2, 2) = C(4, 2) = 6$ ways.

Generalisation: The number of r-multisets that may be chosen from a set of n objects is $C(n - 1 + r, r)$.

The choice of an r-multiset is equivalent to placing r X's into n boxes labelled $a_1, \cdots a_n$ as in Figure 9.4; and that is equivalent to arranging r X's and $(n - 1)$ S's as a 'word'. The word is determined precisely by the choice of positions for the r X's; and that choice can be made in $C(n - 1 + r, r)$ ways.

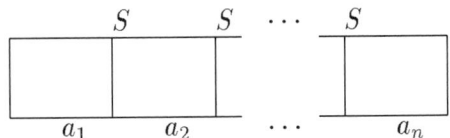

Figure 9.4: r-multisets from $\{a_1, a_2, \cdots a_n\}$

Example: Signals are made by flying five flags on a single pole. Flags of three colours are available in sufficient quantity to have as many flags as required of any colour. The colours are always chosen in the same order (so order is not important). How many signals are possible?

The question asks for the number of 5-multisets from a set of three objects. The solution is:

$$C(3 - 1 + 5, 5) = 21$$

The expression $n(n-1) \cdots 3 \times 2 \times 1$ appears frequently in combinatorial problems and is defined as a function (see section 6.1.1).

Definition 9.1 (Factorial)

$$factorial: \quad N \longrightarrow N$$

$$factorial(n) \; = \; \begin{cases} 1 & \text{if } n = 0 \\ n \times factorial(n-1) & \text{otherwise} \end{cases}$$

We write $factorial(n) = n!$.

Example: Compute 6!

$$
\begin{aligned}
6! = 6 \times 5! \; &= \; 6 \times 5 \times 4! \\
&= \; \cdots \\
&= \; 6 \times 5 \times 4 \times 3 \times 2 \times 1 \times 1 \\
&= \; 720
\end{aligned}
$$

Empty subsets

The above investigation of the counting model, assumes implicitly that r, n are integers and that $1 \leq r \leq n$. The case $r = 0$ should now be considered. A 0-subset would be an empty set. The number of empty subsets of $\{a_1, \cdots a_n\}$ is just 1. So we would like $C(n, 0) = 1$. It is usual for mathematicians to accept similarly that there is one 0-sequence, one 0-permutation and one 0-multiset for any set $\{a_1, \cdots a_n\}$.

Now the results of this section can be stated formally as theorems, whose proofs have mostly been argued above.

Theorem 9.7 *For integers $r, n \geq 0$, the number of r-sequences that can be made from a set of n objects is n^r.*

Proof: The case $r = 0$. $n^0 = 1$, gives one 0-sequence, which agrees with our assumption.

The case $r \geq 1$. The argument is given in problem 9.1 (p. 195).

Theorem 9.8 *The number of r-permutations that can be made from a set of n objects is:*

$$P(n, r) = \begin{cases} n(n-1) \cdots (n-r+1) & \textit{(i.e. } r \textit{ factors)} & \textit{if } n \geq r \geq 1 \\ 1 & & \textit{if } r = 0 \end{cases}$$

Using factorials the theorem may be stated:

$$P(n,r) = \frac{n!}{(n-r)!} \quad \text{where } n \geq r \geq 0.$$

Theorem 9.9 *The number of r-subsets (i.e. r-combinations) from a set of n objects where $0 \leq r \leq n$ is given by:*

$$C(n,r) = \frac{P(n,r)}{r!} = \frac{n!}{r!(n-r)!}$$

Proof: Case $r = 0$. $C(n,0) = 1$ which agrees with our assumption that there is exactly one 0-subset for any set.

The case $r \geq 0$ was argued in problem 9.3 (p. 196).

Theorem 9.10 *The number of r-multisets that can be made from a set of n objects, where $0 \leq r \leq n$ is:*

$$\begin{cases} C(n-1+r,r) & \text{if } n \geq 1 \\ 1 & \text{if } n = 0 \end{cases}$$

Proof: Case $r = 0$. Either $C(n-1+0,0) = 1$ or if $n = 0$ the number of 0-multisets is stated to be 1. So the theorem agrees with our assumption that there is exactly one 0-multiset for any set.

The case $r \geq 0$ was argued in problem 9.4 (p. 196).

9.2.2 Combinatorial reasoning

Theorems 9.11–9.13, below, may be proved algebraically. However each proof is given as an argument in words to illustrate that kind of combinatorial reasoning.

Theorem 9.11 *For integers n, r such that $0 \leq r \leq n$:*

$$C(n,r) = C(n,n-r)$$

Proof: Consider the set of n objects from which the subsets are to be chosen as a universal set.

Then for each subset of r elements there corresponds exactly one complement subset of $n - r$ elements. There is a one-one correspondence between the r-subsets and their complements, which are all the $(n-r)$-subsets. So the number of each is the same. And that is what is stated in the proposition, for $C(n,r)$ counts all the r-subsets and $C(n,n-r)$ counts all the $(n-r)$-subsets.

Theorem 9.12 *For all integers $n \geq 0$:*

$$C(n,0) + C(n,1) + \cdots + C(n,n) = 2^n$$

Proof: Consider a set of $S = \{a_1, a_2, \cdots a_n\}$.

The number of i-subsets of S is $C(n,i)$ and the left-hand side adds the number of 0-subsets, 1-subsets, and so on, up to n-subsets; that is the left-hand side gives the number of all subsets of all possible sizes. Another way to count the subsets is to associate them with n-digit binary strings, by the correspondence 'x_i is in the subset if the ith digit of the binary string is 1, and x_i is not in the subset if the ith digit is 0'. Two choices for each binary digit gives 2^n binary strings and thus 2^n subsets.

$$\text{Therefore,} \quad \sum_{i=0}^{n} C(n,i) = 2^n$$

Theorem 9.13 *For all integers n, k such that $0 \leq k \leq n$:*

$$C(n,k) = C(n-1,k) + C(n-1,k-1)$$

Proof: Consider a set $U = \{a_1, a_2, \cdots, a_s, \cdots a_n\}$ in which 'a_s' is a special element.

All the subsets with k elements (the k-subsets) may be partitioned into two classes: those that contain a_s and those that do not.

The number of all k-subsets is $C(n,k)$.

The number of k-subsets that do not contain a_s is the number of ways of choosing k elements from $U - \{a_s\}$, that is $C(n-1,k)$.

The number of k-subsets that contain a_s is the number of ways of choosing the other $k-1$ elements from $U - \{a_s\}$, that is $C(n-1,k-1)$.

Therefore $C(n,k) = C(n-1,k) + C(n-1,k-1)$.

Together with $C(n,0) = 1$ and $C(n,n) = 1$ the formula for $C(n,k)$ given in theorem 9.13 gives a recursive scheme for the evaluation of $C(n,k)$.

The following examples deal with problems that superficially appear to have nothing in common. However each may be translated into a problem with similar structure to that in the first example. The identification of underlying structural similarity is another aspect of combinatorial reasoning. The problems are all of the type 'sequences with repetition'.

Example 1: How many different eight-letter words can be made with five A's, two B's and one C?

Suppose first that the eight letters are all different, indicated by giving the letters temporary subscripts.

$$A_1 A_2 A_3 A_4 A_5 B_1 B_2 C$$

Under the supposition that the letters are distinct, there are 8! different arrangements. But the A's contribute 5! arrangements for each arrangement of the other letters; and the B's contribute 3! arrangements for each arrangement of the other letters. For completeness note that the C contributes 1! arrangement for each arrangement of the other letters. On removing the subscripts we have left

$$\frac{8!}{5!2!1!}$$

arrangements of the letters, each representing a distinct word.

An alternative way to express the answer to example 1 is:

$$C(8,5) \times C(3,2).$$

This form of the answer is produced by arguing that the A's may be placed in five of the eight available positions in $C(8,5)$ ways. The B's may then be placed in the remaining three places in $C(3,2)$ ways. There is now only one way to place the third letter. Thus the number of ways is $C(8,5) \times C(3,2) \times 1$.

Example 2: How many different eight-digit binary strings have five 0's and three 1's?

In this question there are eight symbols, five of one kind and three of another kind. The question asks us to find all the 'words' of eight letters that can be made from the symbols. As in example 1 the symbols may be given temporary subscripts.

$$0_1 0_2 0_3 0_4 0_5 1_1 1_2 1_3$$

Under the supposition that the symbols are distinct there are 8! different arrangements. But the 0's contribute 5! arrangements for each arrangement of the 1's, and the 1's contribute 3! arrangements for each arrangement of the 0's. Therefore there are

$$\frac{8!}{5!3!}$$

different binary strings.

An alternative way to express the answer to example 2 is: $C(8,5)$. This form of the answer is produced by arguing that the 0's may be placed in five of the eight available positions in $C(8,5)$ ways. The 1's may then be placed in the remaining three places in one way.

Example 3:   Find the number of solutions of the equation

$$x + y + z = 10$$

where x, y, z are integers and $x, y, z \geq 0$. The solutions may be generated by placing ten balls into three boxes labelled x, y, z (Fig. 9.5).

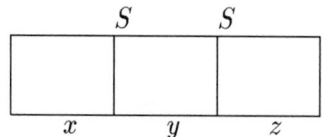

Figure 9.5: Number of solutions

Represent the balls by the letter B and the walls between the boxes by the letter S.

Then each word made from the 12 symbols represents a solution of the equation.

For example $BBSBBBSBBBBB$ represents $x = 2, y = 3, z = 5$ and $BBBBSSBBBBBB$ represents $x = 4, y = 0, z = 6$. The argument of Example 1 shows that the number of solutions of the equation is

$$\frac{12!}{10!2!}.$$

Example 4:   How many numbers from 1 to 10 000 have a digit sum of 7?

Solution: The problem is equivalent to placing seven balls into four boxes (Fig. 9.6).

As in Example 3 let the balls be represented by the letter B and the walls between boxes be represented by the letter S. Each 'word' from the ten symbols represents a solution to the question.

For example $BBSBBBBBSBS$ represents 2410;
                  $SSBBBBBBBSB$ represents 61.

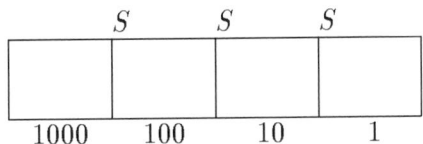

Figure 9.6: Numbers from 1 to 10 000

The argument of Example 1 gives:

$$\frac{10!}{7!3!} = \frac{10 \times 9 \times 8}{3 \times 2 \times 1} = 10 \times 3 \times 4 = 120 \text{ words}$$

There are 120 numbers from 1 to 10 000 with a digit sum of 7.

Example 5:   A computer allocates jobs to four printers. Given 12 jobs to be distributed so that Printer P has 5 jobs, Printer Q has 3 jobs, Printer R has 2 jobs and Printer S has 2 jobs, find the number of ways that the distribution can be made.

Solution: The problem may be regarded as 'the number of words that can be made from the letters PPPPPQQQRRSS'. The answer is

$$\frac{12!}{5!3!2!2!}.$$

9.3   THE BINOMIAL & MULTINOMIAL THEOREMS

9.3.1   The binomial theorem

In this section we consider the expansion of expressions like $(a + b)^n$ where n is a positive integer, using a combinatorial approach rather than algebra.

Let us see what we can learn from the expansion of $(a + b)^3$.

$$
\begin{aligned}
(a + b)^3 &= (a + b)(a + b)(a + b) & (9.1)\\
&= a^3 + 3a^2b + 3ab^2 + b^3 & (9.2)\\
&= C(3,0)a^3b^0 & (9.3)\\
&\quad + C(3,1)a^2b^1 & (9.4)\\
&\quad\quad + C(3,2)a^1b^2 & (9.5)\\
&\quad\quad\quad + C(3,3)a^0b^3 & (9.6)
\end{aligned}
$$

Line 9.1 states what is meant by the symbols on the left-hand side of the equation.

Each term in line 9.2 is constructed by choosing a from m brackets and b from n brackets. A letter must be chosen from each bracket so the terms are of the form

$$a^m b^k \quad \text{where } m + n = 3.$$

The term $3a^2b$ is constructed by choosing a from two brackets and b from one bracket. The selection of a bracket from which to draw b may be done in $C(3,1) = 3$ ways, and the choice of brackets from which to draw a are then automatically the two left.

The term $3a^2b$ corresponds to $C(3,1)a^2b^1$ (line 9.4).

The term a^3 is constructed by choosing a from three brackets and b from none. The term a^3 corresponds to $C(3,0)a^3b^0$ (line 9.3).

In general the terms are of the form $C(3,k)a^m b^k$ where $m + k = 3$.

The result may be stated concisely as:

$$(a + b)^3 = \sum_{k=0}^{3} C(3,k)a^m b^k; \quad m + k = 3$$

$$\text{or} \quad (a + b)^3 = \sum_{k=0}^{3} C(3,k)a^{3-k}b^k$$

The argument may be generalised to any positive integer power by mathematical induction, giving the following theorem.

Theorem 9.14 (Binomial) *For all positive integers n:*

$$
\begin{aligned}
(a + b)^n &= (a + b)(a + b) \cdots (a + b) \\
&= C(n,0)a^n b^0 + C(n,1)a^{n-1}b^1 + \cdots + C(n,n)a^0 b^n \\
&= \sum_{k=0}^{n} C(n,k)a^m b^k; \quad m + k = n \\
&= \sum_{k=0}^{n} C(n,k)a^{n-k}b^k
\end{aligned}
$$

Example 1: Write down the expansion of $(x + \frac{1}{x})^5$.

   Solution:

$$
\begin{aligned}
(x + \frac{1}{x})^5 &= x^5 + C(5,1)x^4(\frac{1}{x}) + C(5,2)x^3(\frac{1}{x})^2 + \cdots \\
&= x^5 + 5x^3 + 10x + 10x^{-1} + 5x^{-3} + x^{-5}
\end{aligned}
$$

Example 2: Write down the coefficient of x^{11} in $(3x + 2x^2)^9$.

Solution:

$$(3x + 2x^2)^9 = \sum_{i=0}^{9} C(9,i)(3x)^{9-i}(2x^2)^i$$

$$= \sum_{i=0}^{9} C(9,i)3^{9-i}2^i x^{9+i}$$

If $9 + i = 11$ then $i = 2$. The coefficient of x^{11} is:

$$C(9,2)3^7 2^2 = \frac{9 \times 8}{2 \times 1} \times 3^7 \times 2^2 = 314\,928$$

Example 3: Prove that $\displaystyle\sum_{k \text{ odd}} C(n,k) = \sum_{k \text{ even}} C(n,k)$.

Solution: Set $a = 1, b = -1$ in theorem 9.14.

Then we obtain:

$$0 = C(n,0) - C(n,1) + C(n,2) - \cdots + (-1)^n C(n,n)$$

Hence the sum of the $C(n,k)$ for k odd equals the sum of the $C(n,k)$ for k even.

9.3.2   The multinomial theorem

The binomial theorem may be generalised in one way by raising a multinomial expression to a positive integer power.

Theorem 9.15 (Multinomial) *For positive integers n:*

$$(x_1 + x_2 + \cdots + x_m)^n$$

is the sum of all terms of the form

$$\frac{n!}{r_1! r_2! \cdots r_m!} x_1^{r_1} x_2^{r_2} \cdots x_m^{r_m}$$

for all non-negative integer solutions of $r_1 + r_2 + \cdots + r_m = n$.

Proof: A term in the expansion is obtained from the product

$$(x_1 + x_2 + \cdots + x_m)(x_1 + x_2 + \cdots + x_m) \cdots (x_1 + x_2 + \cdots + x_m)$$

by selecting one term from each of the n brackets.

The term in
$$x_1^{r_1} x_2^{r_2} \cdots x_m^{r_m}$$
is obtained by selecting x_1 from r_1 brackets, x_2 from r_2 brackets and so on.

Each selection corresponds to a 'word' constructed from r_1 letters x_1, r_2 letters x_2 and so on. The selection can be done in
$$\frac{n!}{r_1! r_2! \cdots r_m!}$$
ways.

Example 1: Expand $(x + y + z)^3$ by the multinomial theorem.

Solution:
$$(x + y + z)^3 = \sum \frac{3!}{p! q! r!} x^p y^q z^r$$
for all integer solutions of $p + q + r = 3; p, q, r \geq 0$.

We obtain
$$\begin{aligned}(x + y + z)^3 &= x^3 + y^3 + z^3 + 3x^2 y + 3xy^2 + 3y^2 z + 3yz^2 + 3x^2 z \\ &\quad + 3xz^2 + 6xyz\end{aligned}$$

There are ten terms in all which is the number of non-negative integer solutions of $p + q + r = 3$.

Example 2: How many terms are there in the expansion of $(x + 2y + 3z)^5$ by the multinomial theorem?

Solution: The problem is equivalent to finding the number of non-negative integer solutions of $p + q + r = 5$.

The answer is $\dfrac{7!}{5! 2!} = 21$.

Example 3: Find the coefficient of x^5 in the expansion of $(1 + 3x + 2x^2)^4$.

The general term of the expansion is:
$$\frac{4!}{p! q! r!} 1^p (3x)^q (2x^2)^r; \quad p + q + r = 4, \quad p, q, r, \geq 0$$
$$= \frac{4!}{p! q! r!} (3)^q 2^r x^{q+2r}$$

The terms in x^5 are given by the integer solutions of
$p + q + r = 4, q + 2r = 5$ satisfying $0 \leq p, q, r \leq 4$.

The solutions are $p = 1, q = 1, r = 2$ and $p = 0, q = 3, r = 1$.

The coefficient of x^5 is $\dfrac{4!}{1!2!} \times 3 \times 2^2 + \dfrac{4!}{3!1!} \times 3^3 \times 2 = 360$.

9.4 ALGORITHMS

The main theme of this chapter is counting. However for some problems it is useful to *generate* the permutations, subsets or other objects being counted. In this section some algorithms are provided for generating subsets and one algorithm for generating permutations.

9.4.1 Subsets of a set

Three algorithms are described to answer the problem 'how to generate all the subsets of a finite set'.

Algorithm 9.1 *To generate all the subsets of a finite set.*

      input: A set $S = \{x_1, x_2, \cdots, x_n\}$
                 n, the number of elements in S.
    output: All the subsets of S.

The idea behind the algorithm is to construct a characteristic function to identify the elements of S for each subset. If S has n elements, the binary strings with n digits serve the purpose. For a given binary string, each digit corresponds to an x in S. If the digit is 1 then include x in the subset; if the digit is 0 then exclude x.

    method: $m \leftarrow 0$
                 While $m < 2^n$:
                       express m as a binary string of length n
                       $(m = a_1 2^{n-1} + a_2 2^{n-2} + \cdots + a_n 2^0)$
                       define the characteristic function
                       $(c_i \leftarrow a_i \text{ for } i = 1, 2, \cdots, n)$
                       choose the subset Y corresponding to m
                       (if $c_i = 1$ then include x_i in Y)
                       record Y (print or store in a file)
                       $m \leftarrow m + 1$

Example 1: Find all the subsets of $S = \{a, b, c\}$ using algorithm 9.1.

m	$c_1\ c_2\ c_3$	Subset Y
0	0 0 0	$\emptyset$
1	0 0 1	$\{c\}$
2	0 1 0	$\{b\}$
3	0 1 1	$\{b,c\}$
4	1 0 0	$\{a\}$
5	1 0 1	$\{a,c\}$
6	1 1 0	$\{a,b\}$
7	1 1 1	$\{a,b,c\}$

Algorithm 9.2 *To generate all the subsets of a finite set.*

method: If the set S is empty, then $\emptyset$ is the only subset of S; record $\emptyset$.

Otherwise select arbitrarily an element $x \in S$.

Find all the subsets of $S - \{x\}$ and

for each subset Y of $S - \{x\}$, record $\{x\} \cup Y$.

The idea behind this algorithm is to use a recursive approach. The statement 'find all the subsets of $(S - \{x\})$' involves a recursive call.

The sets $S, S - \{x\}, \cdots$ form a strictly decreasing sequence, for each set is a proper subset of the one before. The sequence must end with the empty set, $\emptyset$, since S is finite.

Hence the recursion will terminate.

Example 2:   Use algorithm 9.2 to find all the subsets of $\{a,b,c\}$.

Define *subset* to be a function from sets to sets such that *subsetX* means output all the subsets of X.

S	x	$S - \{x\}$	Y	*output*
$\{a,b,c\}$	a	$\{b,c\}$	$subset\{b,c\}$ (defer (1))	
$\{b,c\}$	b	$\{c\}$	$subset\{c\}$ (defer (2))	
$\{c\}$	c	$\{\ \}$	$subset(\emptyset)$	$\emptyset$
			$\emptyset$	$\{c\}$
return(2)			$\emptyset, \{c\}$	$\{b\}, \{b,c\}$
return(1)			$\emptyset, \{c\}\{b\}, \{b,c\}$	$\{a\}, \{a,c\},$
				$\{a,b\}, \{a,b,c\}$

Observe that algorithm 9.1 and algorithm 9.2 generate subsets in the same order.

The idea behind the next algorithm is to generate the subsets in a particular order. The requirement is that successive subsets should differ by exactly one element. The algorithm uses what is called a Gray code.

Definition 9.2 *A Gray code of length n is a sequence of 2^n binary strings, each of length n, and each of which differs from its neighbours in the sequence by exactly one digit.*

Example 3: A Gray code of length 2 is the sequence $11, 01, 00, 10$.

The Gray code in the example corresponds to a Hamiltonian cycle on a 4-vertex graph, as shown in Figure 9.7.

A recursive algorithm for constructing Gray codes is given below in algorithm 9.3.

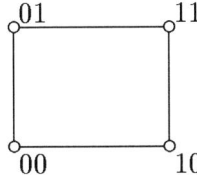

Figure 9.7: A 4-vertex graph

Algorithm 9.3 (Gray code) *To generate all the binary sequences of length k in order, so that successive sequences differ by exactly one binary digit.*

method: Start with the sequence of all zeros. To pass from the sequence

$$b_1 b_2 b_3 \cdots b_k$$

to the next sequence we examine the digit sums

$$b_1 + b_2 + b_3 + \cdots + b_k,$$
$$b_2 + b_3 + \cdots + b_k,$$
$$b_3 + \cdots + b_k, \cdots \text{etc.}$$

If $b_i + b_{i+1} + \cdots + b_k$ is the *first* of these sums to be *even*, we replace the digit b_i by 0 if it is 1 or 1 if it is 0.

If *all* the sums are odd it means that we have reached the last sequence.

Example 4: Find a Gray code of length 3.

The Gray code generated by algorithm 9.3 with $k = 3$ is

$$000, 100, 110, 010, 011, 111, 101, 001$$

A Gray code of length 3 corresponds to a Hamiltonian circuit on an 8-vertex graph (i.e. a cube).

Algorithm 9.4 *To generate all the subsets of a finite set using a Gray code.*

    input: A set $S = \{x_1, x_2, \cdots, x_n\}$
               n, the number of elements in S.
    output: All the subsets of S.
               Postcondition: the subsets generated should be recorded in an order such that successive sets differ by exactly one element.

    method: Construct a Gray code of length n.
               Select subsets of S using the Gray code as characteristic function.

Example 5: Find all the subsets of $\{a, b, c\}$ using algorithm 9.4.

    Solution: Using the Gray code generated in example 4 as characteristic function, subsets are obtained in the following order.

$$\emptyset, \{a\}, \{a, b\}, \{b\}, \{b, c\}, \{a, b, c\}, \{a, c\}, \{c\}$$

9.4.2 Permutations

Many algorithms for generating permutations are known. The algorithm presented here produces all n-permutations of a set of n objects, $A = \{a_1, a_2, \cdots, a_n\}$.

In section 9.2.1 a permutation of the elements of a set was described as an 'arrangement' of those elements in which order was important and repetition was not allowed. The description is not satisfactory as a definition because a 'permutation' is said to be an 'arrangement'. That is, one word has been explained by another word. The following definition of permutation overcomes the problem by defining a permutation as a special kind of function.

Definition 9.3 (Permutation) *A permutation P of a finite set A is a one-one function from A onto A.*

Example 1: The function σ is a one-one mapping of

$$A = \{a_1, a_2, a_3, a_4, a_5\}$$

onto A, defined by

$$\sigma(a_1) = a_2, \sigma(a_2) = a_1, \sigma(a_3) = a_5, \sigma(a_4) = a_2, \sigma(a_5) = a_3.$$

Because the function σ is one-one from A onto A it is a permutation of A.

The permutation σ may be defined by the array notation

$$\begin{pmatrix} a_1 & a_2 & a_3 & a_4 & a_5 \\ a_4 & a_1 & a_5 & a_2 & a_3 \end{pmatrix}$$

in which each element on the top line maps to the element directly below. If the original order $a_1 a_2 a_3 a_4 a_5$ is known then the permutation is sufficiently identified by $(a_4, a_1, a_5, a_2, a_3)$ or merely the subscripts 41523.

A measure of the extent to which each element is 'out of order' relative to the 'natural order' (12345) of the sequence is given by a *displacement* function d.

The value of the displacement function at a_i, namely $d(a_i)$, is the number of elements of A to the left of a_i in the permutation that would be to the right of a_i in the natural order. In the example above $d(a_2) = 2$ because a_4, a_5 are to the left of a_2 in the permutation but to the right of a_2 in the natural order. Similarly $d(a_1) = 1, d(a_3) = 2, d(a_4) = 0$, and $d(a_5) = 0$.

The *displacement* function d is defined informally

$$d : \quad \{a_1, a_2, \cdots, a_n\} \longrightarrow \{0, 1, \cdots, (n-1)\}$$

$$d(a_i) = d_i = \begin{cases} 0 & \text{if } a_i \text{ in permutation } \sigma \text{ is mapped to itself, or to} \\ & \text{an element to its left in the original order} \\ k & \text{if } a_i \text{ is mapped to an element } k \text{ places to the} \\ & \text{right in the original order.} \end{cases}$$

Example 2: Find the displacement sequence $(d_1, d_2, \cdots, d_5)$ for the permutation σ defined in Example 1 above.

Solution:

	a_1	a_2	a_3	a_4	a_5
	a_4	a_1	a_5	a_2	a_3
i	1	2	3	4	5
$d(i)$	1	2	2	0	0

The displacement sequence is $(1, 2, 2, 0, 0)$.

Definition 9.4 (Displacement) *The* displacement function *for a permutation σ is the function*

$$d: \quad \{a_1, a_2, \cdots, a_n\} \longrightarrow \{0, 1, \cdots, (n-1)\}$$

$$\text{let } \sigma^{-1}(a_i) = a_j, \quad \text{then:}$$

$$d(a_i) = d_i = \begin{cases} 0 & \text{if } j \leq i \\ j - i & \text{if } j > i \end{cases}$$

The sequence $d_1, d_2, \cdots, d_n$ is called the displacement sequence *for σ.*

Theorem 9.16 *The displacement function d satisfies the inequality*

$$0 \leq d_i \leq n - i$$

for $i = 1, 2, \cdots, n$.

Proof: By induction on $(n - i)$.

   Basic step: If $n - i = 0$ then $i = n$,

   and the last element cannot be displaced to the right, so $d(n) = 0$;
   and $0 \leq d(n) \leq 0$ implies $d(n) = 0$; so the proposition is true for
   $n - i = 0$.

   Inductive step: Suppose the proposition is true for $n - i = k$.
   Then $i = n - k$ and $0 \leq d(n - k) \leq k$. Now consider $n - i = k + 1$.
   This gives $i = n - k - 1$, and there is one more position available on
   the right of a_i for displacing this element. Therefore the proposition
   is also true for $n - i = k + 1$.

The converse problem to finding a displacement sequence for a given permutation is to find the permutation corresponding to a given displacement
sequence.

Algorithm 9.5 *To generate a permutation from a displacement sequence.*

   Informal description : Given a sequence of n objects in natural
   order

$$a_1, a_2, \cdots, a_n$$

   n vacant positions, and a displacement sequence

$$d_1, d_2, \cdots, d_n \quad (0 \leq d_i \leq n - i))$$

   place a_1 in position $d_1 + 1$ (counted from the left).
   If $a_1, a_2, \cdots, a_{i-1}$ have already been placed, put a_i in the $(d_i + 1)$th
   vacant position (counted from the left).

Example 3: Find the permutation corresponding to the displacement sequence $3, 3, 0, 1, 1, 0$.

1. Place a_1 in position $(3 + 1)$.

$-, -, -, a_1, -, -$

2. Place a_2 in the $(3 + 1)$th vacant position.

$-, -, -, a_1, a_2, -$

3. Place a_3 in the $(0 + 1)$th vacant position.

$a_3, -, -, a_1, a_2, -$

4. Place a_4 in the $(1 + 1)$th vacant position.

$a_3, -, a_4, a_1, a_2, -$

5. Place a_5 in the $(1 + 1)$th vacant position.

$a_3, -, a_4, a_1, a_2, a_5$

6. Place a_6 in the $(0 + 1)$th vacant position.

$a_3, a_6, a_4, a_1, a_2, a_5$

Therefore the permutation is $a_3, a_6, a_4, a_1, a_2, a_5$.

Theorem 9.17 *There is a one-one correspondence between the set of all permutations on a set A with n elements and the set of all displacement sequences with n terms.*

Proof: The number of permutations on a set A with n members is $n!$.

The number of distinct displacement sequences with n terms is

$$n \times (n - 1) \times \cdots \times 2 \times 1 = n!$$

from the previous theorem and the multiplication rule for counting.

Exactly one permutation on a set of n objects can be generated from each valid displacement sequence with n terms. See algorithm 9.5. The result follows.

Algorithm 9.6 *To generate a permutation from a displacement sequence.*

Formal version:

input: A displacement sequence $(d_1, d_2, \cdots, d_n)$.
A permutation $(a_1, a_2, \cdots a_n)$
Precondition: the displacement sequence and the permutation each have n members, and satisfy their respective definitions.

output: A permutation
$$\begin{pmatrix} a_1 & a_2 & \cdots & a_n \\ p_1 & p_2 & \cdots & p_n \end{pmatrix}$$
Postcondition: the permutation $(p_1, \cdots, p_n)$ matches the input displacement sequence.

The steps in the algorithm are first described in words.

method: Label the n 'positions' in the permutation from 1 to n.
For $i = 1$ to n:
place a_i in the position labelled $d_i + 1$
relabel the positions following position $d_i + 1$

The steps in the algorithm are now described again in more detail.

method: For $k = 1$ to n
$c(k) = k$
For $i = 1$ to n
$k \leftarrow d_i + 1$
$j \leftarrow c(k)$
$p_j \leftarrow a_i$
for $m = k$ to $n - i$
$c(m) \leftarrow c(m + 1)$

Example 4: Find the permutation of $(a_1, a_2, \cdots a_6)$ corresponding to the displacement sequence $(3, 3, 0, 1, 1, 0)$.

Let the permutation be

$$\begin{pmatrix} a_1 & a_2 & a_3 & a_4 & a_5 & a_6 \\ p_1 & p_2 & p_3 & p_4 & p_5 & p_6 \end{pmatrix}$$

The following table shows successive domain values for c through the running of the algorithm.

k						i
1	2	3	4	5	6	1
1	2	3	-	4	5	2
1	2	3	-	-	4	3
-	1	2	-	-	3	4
-	1	-	-	-	2	5
-	1	-	-	-	-	6
$c(k) =$ 1	2	3	4	5	6	

Stepping through the algorithm produces the following table of values for the variables.

i	d_i	k	$c(k)$	j	output
1	3	4	4	4	$p_4 = a_1$
2	3	4	5	5	$p_5 = a_2$
3	0	1	1	1	$p_1 = a_3$
4	1	2	3	3	$p_3 = a_4$
5	1	2	6	6	$p_6 = a_5$
6	0	1	2	2	$p_2 = a_6$

The output is the permutation

$$\begin{pmatrix} a_1 & a_2 & a_3 & a_4 & a_5 & a_6 \\ a_3 & a_6 & a_4 & a_1 & a_2 & a_5 \end{pmatrix}$$

Example 5: Find the permutation of $(a_1, a_2, \cdots a_6)$ corresponding to the displacement sequence $(0, 2, 4, 0, 1, 0)$

Solution: The displacement sequence is illegal since $d_3 = 4$ contradicts Theorem 9.16 which states that $d_3 \leq 6 - 3$.

Algorithm 9.7 (Permutations) *To generate all n-permutations of a set of n objects.*

> **input:** A set $\{a_1, a_2, \cdots a_n\}$.
> **output:** All permutations of A.
> **method:** $D \leftarrow (0, 0, \cdots, 0)$
> *(the n-element, null displacement sequence).*
> Repeat:
>> find a new displacement sequence (algorithm 9.8)
>> convert the sequence to a permutation (algorithm 9.6)
>> record the permutation
> until $n!$ permutations have been recorded

Algorithm 9.8 *To find a new displacement sequence from a given displacement sequence.*

> **input:**   A displacement sequence $D = (d_1, d_2, \cdots d_n)$.
> **output:** The next displacement sequence in 'dictionary' order, D'.
> **method:** $i \leftarrow n - 1$
> While $d_i = n - i$ and $i \neq 0$:
> $\qquad d_i \leftarrow 0$
> $\qquad i \leftarrow i - 1$
> if $i > 0$ then $d_i \leftarrow d_i + 1$

Example 6: Given the permutation $(a_5, a_2, a_4, a_3, a_1)$ generate the next permutation using algorithms 9.8 and 9.6.

Solution: Permutation = 52431. Displacement sequence = 41210.

Next displacement sequence = 42000. The new permutation is 34251.

Algorithm 9.7 may be used to solve question 7.4.10. That problem is of the 'travelling salesman' type, and may be solved exactly by generating all sequences of five jobs, and calculating the cost for each job sequence.

The algorithm generates 120 job schedules and their costs. For example 12345 costs $33, \cdots, 14325$ costs $28, \cdots, 43125$ costs 28, and, finally 54321 costs 42.

9.5   EXERCISES

1. How many numbers divisible by 7 are there in the set of consecutive integers from 100 to 999?

2. How many members has the sequence

$$a_{\lfloor \frac{n}{2} \rfloor}, \cdots, a_n \quad \text{where } n \geq 2?$$

   Consider the case n even and the case n odd, separately.

3. What is the 27th odd number after 99?

4. Given the set

$$S = \{ j \in Z \mid 100 \leq j \leq 999 \}$$

   (a) Find $|S|$.

   (b) How many odd integers are in S?

   (c) How many integers in S have distinct digits?

(d) How many odd integers in S have distinct digits?

5. A PC system has available two kinds of hard disk, three types of VDU screen, two keyboards and three printer types. How many distinct systems can be purchased?

6. The number 42 has the prime factorisation $2 * 3 * 7$. Thus 42 may be written in four ways as the product of two positive integer factors.

   (a) In how many ways can 24 be written as the product of two positive integer factors?

   (b) The prime factorisation of a positive integer n is $p_1 * p_2 * \cdots * p_k$. Given that the prime factors are distinct, in how many ways can n be written as the product of two positive integer factors?

7. How many integers from 1 to 1000 are not divisible by 8 or 12?

8. How many integers from 1 to 1000 are divisible by 6 or 8?

9. Let S be the set of four-digit numbers made from the set of digits $\{0, 2, 3, 5, 6\}$ allowing repetitions.

   (a) How many numbers are in S?

   (b) How many numbers in S have no repeated digit?

   (c) How many of those in part (b) are even?

   (d) How many of those in part (b) are greater than 4000?

   (e) How many of those in part (b) are exactly divisible by 5?

10. Each character in Morse Code is represented by a sequence of up to four dots or dashes. For example 'b' is represented by $-\cdots$ and 's' is represented by $\cdots$
How many characters can be represented in this way?

11. A *complete* graph is defined to be a simple graph in which every pair of distinct vertices is adjacent, that is connected by an edge. The symbol for a complete graph with n vertices is K_n. Find the number of edges in K_n.

12. The complete *bipartite* graph $K_{m,n}$ is a simple graph G in which

$$V(G) = A \cup B, \ A \cap B = \emptyset, \ |A| = m, |B| = n, \text{ and}$$

$$E(G) = \{\{a,b\} \mid a \in A, b \in B\}$$

Calculate the number of edges in $K_{m,n}$.

13. How many integers from 1 to 1000 inclusive contain a 5 or a 6 in their decimal representation?

14. How many integers from 1 to 1000 inclusive are divisible by none of 5, 6 or 8?

15. Six people are introduced in a tutorial. Use the pigeonhole principle to show that at least three have all met before or at least three have not met at all. (*Hint:* Draw a graph.)

16. A wheel is divided into 24 sectors, each numbered from 1 to 24 but arranged at random. Prove that there are four consecutive numbers with sum greater than 50.

17. How many permutations can be made using all the letters of each word?
(a) WAGGA   (b) MURRUMBURRAH   (c) COONABARABRAN

18. The winner of a competition is to be decided by a panel of three judges. First the best ten entries are selected. Then each of the judges chooses a winner from the ten best entries. For the highest prize to be awarded an entry must be selected as winner by at least two of the judges. Calculate:

   (a) the number of different possible selections of winners by the three judges.

   (b) the number of selections in which the highest prize is awarded.

   (c) the fraction of possibilities in which the highest prize is awarded.

19. How many integers from 1 to 10 000 have a digit sum of 9?

20. How many different 10-digit binary strings have six 0's and four 1's?

21. Find the coefficient of $x^5 y^4 z^3$ in $(x - 2y + 3z + w)^{12}$.

22. How many solutions has $x + y + z = 15$, where x, y, z are non-negative integers?
Find the number of terms in the expansion of $(u + v + w)^{15}$.

23. How many 12 digit binary strings have eight 0's and four 1's?

24. How many books must be chosen from among 24 mathematics books, 25 computer science books, 21 literature books, 15 economics books and 30 accounting books to ensure that there are at least 12 books on the same subject?

25. From a 'short' list of 4 men and 6 women three branch managers are to be chosen. How many different selections can be made if at least one person of each sex is to be chosen?

26. Prove that if 151 integers are selected from $\{1, 2, 3, \cdots 300\}$ then the selection must include two integers a, b where a divides b or b divides a.

27. Generate all the subsets of $\{a, b, c, d\}$.

28. Find the permutation generated immediately after $(a_5, a_4, a_3, a_1, a_2)$ by algorithm 9.7.

Chapter 10

Algebraic Structures

There are two main objectives in this chapter. The first is to identify algebraic structures and algebraic properties that appear frequently in computing, and by providing a language to describe those structures to simplify and help our understanding of some of the complex systems involved. The second objective is to prove some general properties about the algebraic structures, which is economical when those structures occur in more than one place.

You should be able to use the language of structures and be able to identify particular algebraic properties; and at a higher level you should be able to prove properties about general structures.

The chapter introduces semigroups, monoids, groups; partially ordered sets, lattices, Boolean algebras and finite fields, with a discussion of elementary properties and an indication of where they appear in computing.

10.1 ALGEBRAS

Early in this book we observed how mathematical systems could be distinguished from each other by a study of the operations defined on them.

Example: The operation of subtraction is defined on the integers and on the natural numbers. However subtraction on the integers is a *total* function whereas subtraction on the natural numbers is not total, only partial. Put another way, subtraction is *closed* on the integers but not on the natural numbers.

There are other structural features by which systems may be shown to be different. For example the natural numbers (N) are 'well-ordered' (i.e. every subset of N has a least element), whereas the real numbers and the integers are not.

In this section certain *algebraic* properties of *operations* are defined followed by a description of some important 'algebras' or 'algebraic structures'. Informally an *algebra* consists of a set of objects together with one or more operations on the set. The properties of the operations characterise the algebra. Sometimes more than one set is involved.

10.1.1   Operations and their algebraic properties

Many operations have been used in this book, for example multiplication and addition on integers, union and intersection on sets in Chapter 3 and 'append' on lists in Chapter 6. Part of the definition of each operation is a declaration of the set on which it is defined. For convenience the following definitions are repeated from Chapter 3.

Definition 10.1 (Binary operation) *A binary operation on a set S is a function from $S \times S$ to S.*

Definition 10.2 (Unary operation) *A unary operation on a set S is a function from S to S.*

In each case the word 'on' implies that the 'values' of the function do not fall outside the set. Another way of putting it is to say that the set is *closed under the operation.*

Example 1: The *addition* function is a binary operation on **Z**. That is, **Z** is closed under addition.

The function
$$addition : \mathbf{Z} \times \mathbf{Z} \to \mathbf{Z}$$

$$addition(x, y) = x + y$$

satisfies the definition for a binary operation on **Z**.

(The arrangement $x + y$ is called *infix* notation; $+(x, y)$ is called *prefix* notation and $(x, y)+$ is called *postfix* notation. Each is in use, and they all represent the same thing: the sum of x and y.)

Example 2: Multiplication is a binary operation on $M_{n \times n}$, the set of square matrices with n rows and n columns.

Multiplication is a function from ordered pairs of matrices of size $n \times n$ to matrices of size $n \times n$.

$$multiplication(A, B) = AB$$

Writing A, B side by side is taken to mean that A and B should be multiplied in that order. (This is called juxtaposition notation.)

Example 3: With respect to a universal set X, 'complement' is a unary operation on $\mathcal{P}(X)$, the power set of X.

The complement is defined by

$$complement: \quad \text{power set of } X \longrightarrow \text{power set of } X$$

$$complement(A) = A' = X - A$$

Example 4: 'Concatenation' of strings is an operation on A^* the set of all strings constructed from the alphabet A see Chapter 3, p. 62.

Let $\alpha = a_1 a_2 \cdots a_r, \beta = b_1 b_2 \cdots b_s$ be strings in A^*, so that each a_i and each b_j is in the alphabet A.

Then *concatenation* is a function from $A^* \times A^*$ to A^* and

$$
\begin{aligned}
concatenate(\alpha, \beta) &= \alpha \text{ followed by } \beta \\
&= \alpha\beta \\
&= a_1 a_2 \cdots a_r b_1 b_2 \cdots b_s
\end{aligned}
$$

Structure properties of binary operations

Although the same words may be used for operations on different sets, the functions themselves are different. The difference is an immediate consequence of the change in domain. It may also happen that the structure properties are different. Consider 'multiplication':

1. For all real numbers x, y:
$$xy = yx$$

2. For some $n \times n$ matrices A, B:

$$AB \neq BA$$

For example $\begin{pmatrix} 1 & 2 \\ -1 & -2 \end{pmatrix} \begin{pmatrix} 1 & 1 \\ 2 & 2 \end{pmatrix} \neq \begin{pmatrix} 1 & 1 \\ 2 & 2 \end{pmatrix} \begin{pmatrix} 1 & 2 \\ -1 & -2 \end{pmatrix}$

In (1), the order of the variables x, y is not important. In (2) the order of the matrices *is* important. We say that 'multiplication' defined on real numbers is *commutative* but 'multiplication' defined on $n \times n$ matrices is not commutative.

The *commutative* property is an algebraic property that may belong to an operation on a set. To generalise the discussion and emphasise the structure properties rather than the specific properties of particular operations, let us denote an *arbitrary* operation on a set S by the symbol $\square$. (So $\square$ stands for any of $*, +, \vee, \wedge, \cup, \cap$, concatenate, or any other operation.) Then the following structure properties are defined.

Definition 10.3 (Commutative) *A binary operation* $\square$ *on S is said to be commutative if and only if*

$$x \square y = y \square x \quad \text{for all } x, y \text{ in } S$$

Example: The *concatenation* operation is not commutative on strings. For if $\alpha = car$ and $\beta = pet$ then $concatenate(\alpha, \beta) = carpet$ and $concatenate(\beta, \alpha) = petcar$.

Definition 10.4 (Associative) *A binary operation* $\square$ *on S is said to be associative if and only if*

$$x \square (y \square z) = (x \square y) \square z \quad \text{for all } x, y, z \text{ in } S$$

If a binary operation $\square$ is known to be associative then there is no ambiguity in writing an expression such as $x \square y \square z$, because it doesn't matter whether this is interpreted as $(x \square y) \square z$ or $x \square (y \square z)$.

Example 1: The operation of composition of functions, denoted by $\circ$, is associative on functions where that composition is defined:

Let f, g, h be functions for which $f(g(h(x)))$ is defined for at least one x. Then:

$$f \circ (g \circ h) = (f \circ g) \circ h$$

The point is that function 'composition' amounts to an application of a function 'followed by' an application of a second function.

To show that function composition is associative, consider an element x for which $h(x) = x', g(x') = x'', f(x'') = x'''$ are all defined.

Then $g \circ h(x) = g(h(x)) = g(x') = x''$

and so $f \circ (g \circ h)(x) = f(g \circ h(x)) = f(x'') = x'''$.

Also $h(x) = x'$ and $(f \circ g) \circ h(x) = (f \circ g)(x') = f(g(x')) = f(x'') = x'''$.

So for all such x's:

$$f \circ (g \circ h)(x) = (f \circ g) \circ h(x)$$

That is, for suitable domains:

$$f \circ (g \circ h) = (f \circ g) \circ h$$

The argument is illustrated in the directed graph of Figure 10.1, in which for example, if f maps x to x' then x, x' are represented by vertices of the graph and the edge (x, x') is labelled f.

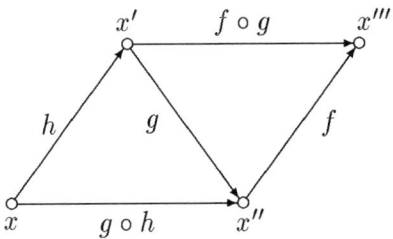

Figure 10.1: Associativity of function composition

Example 2: Matrix multiplication is associative.

The best way of showing that matrix multiplication is associative belongs to a course in linear algebra, where we would learn that matrices correspond to certain kinds of functions (called 'linear transformations') and hence matrix multiplication is associative by example 1.

Example 3: Subtraction on the set of integers is *not* associative.

The counterexample $5 - (3 - 2) = 4$ and $(5 - 3) - 2 = 0$ shows that it is not true that

$$x - (y - z) = (x - y) - z \text{ for all } x, y, z \in \mathbf{Z}.$$

Thus subtraction is not associative on integers.

Definition 10.5 (Idempotent) *A binary operation $\square$ on S is said to be idempotent if and only if*

$$x \square x = x \quad \text{for all } x \text{ in } S$$

Example 1: Intersection is idempotent.

$$X \cap X = X \quad \text{for all sets } X$$

Example 2: Multiplication on the set of integers is not idempotent.

The only solutions of $x \times x = x$ are $x = 0$ and $x = 1$.

For multiplication to be idempotent on numbers it would be necessary to restrict the set of numbers to $\{0, 1\}$.

Definition 10.6 (Identity) *An identity for the binary operation $\square$ on S is an element 'e', say, in S for which*

$$x \square e = x = e \square x \quad \text{for all } x \text{ in } S.$$

An element e is called an 'identity for $\square$ in S'. A property of an identity is that if it exists then it is unique.

Theorem 10.1 *If there is an identity 'e' for an operation $\square$ on a set S then that identity is unique.*

Proof: Let e, f be identities for $\square$. Then:

$$e = e \square f = f$$

That is $e = f$. So there is at most one identity.

From now on it is legitimate to speak of *the* identity for $\square$ on S.

Example 1: The identity for multiplication on $M_{2 \times 2}$ is

$$\mathbf{I} = \begin{pmatrix} 1 & 0 \\ 0 & 1 \end{pmatrix}.$$

By the definition of matrix multiplication we have:

$$\begin{pmatrix} a & b \\ c & d \end{pmatrix} \begin{pmatrix} 1 & 0 \\ 0 & 1 \end{pmatrix} = \begin{pmatrix} a & b \\ c & d \end{pmatrix} = \begin{pmatrix} 1 & 0 \\ 0 & 1 \end{pmatrix} \begin{pmatrix} a & b \\ c & d \end{pmatrix}.$$

Example 2: The identity for addition on R is 0, since:

$$\forall x \in R \quad x + 0 = x = 0 + x.$$

Example 3: The identity for union ($\cup$) on the power set of A is the empty set, $\emptyset$.

We have for all subsets X of A (i.e. $\forall X \in \mathcal{P}(A)$):

$$X \cup \emptyset = X = \emptyset \cup X.$$

Example 4: The identity for concatenation is the 'empty string' λ.

$concatenate(\alpha, \lambda) = \alpha = concatenate(\lambda, \alpha)$.

If the concept of the 'empty string' is not allowed then there is no identity for concatenation on strings.

Definition 10.7 (Inverse) *Let e be the identity for* $\Box$ *on S. The element* x^{-1} *is an* inverse *of x with respect to* $\Box$ *on S if and only if*

$$x \Box x^{-1} = e = x^{-1} \Box x$$

Note that each x has its own 'inverse'. If $\Box$ represents an 'additive' type of operation it is usual to denote the inverse of x by $(-x)$.

Theorem 10.2 *Let the binary operation* $\Box$ *on S be associative. Let there be an identity element e (unique by theorem 10.1). If a is an inverse of x with respect to* $\Box$ *then the inverse a is unique (for each x).*

Proof: Suppose that a, b are both inverses of x with respect to $\Box$ on S. Let e be the identity for $\Box$ on S. Then:

$$
\begin{aligned}
a &= a \Box e && \text{since } e \text{ is the identity} \\
&= a \Box (x \Box b) && \text{since } b \text{ is an inverse for } x \\
&= (a \Box x) \Box b && \text{since } \Box \text{ is associative} \\
&= e \Box b && \text{since } a \text{ is an inverse for } x \\
&= b && \text{since } e \text{ is the identity}
\end{aligned}
$$

So there is at most one inverse of x with respect to $\Box$ on S.

Theorem 10.3 *Let a be the inverse of x with respect to the operation* $\Box$ *on the set S. Then x is the inverse of a. That is* $(x^{-1})^{-1} = x$.

Proof: Let e be the identity for the binary operation $\Box$ on S and let a be the inverse of x with respect to $\Box$.

Then $x \Box a = e = a \Box x$ (definition 10.7)

But equality is symmetric.

Thus $a \Box x = e = x \Box a$

So by definition 10.7, x is the inverse of a.

Example 1: There is no inverse of 0 with respect to multiplication($\times$) on the real numbers.

The identity for multiplication is 1. But the equation $0 \times x = 1$ has no solution for x and so there is no inverse of 0 with respect to multiplication.

Example 2: For each element $a \in (R - \{0\})$ there is an inverse of a with respect to multiplication. The identity for multiplication is 1. For each non-zero real number a there is a real number $\frac{1}{a}$ such that

$$a \times \frac{1}{a} = 1 = \frac{1}{a} \times a$$

and so $\frac{1}{a}$ is the inverse of a for each $a \in (R - \{0\})$.

Example 3: There is an inverse of each real number with respect to addition on the real numbers.

The identity for addition is 0 and for each real number a there is a real number $(-a)$ such that

$$a + (-a) = 0 = (-a) + a.$$

So $(-a)$ is the inverse of a with respect to addition for each real number a.

Example 4: There is no inverse of a non-empty string with respect to concatenation.

The identity is λ, the empty string. Consider a string $\alpha = a_1 a_2 \neq \lambda$

$$concatenate(a_1 a_2, x) = a_1 a_2 x$$

and $a_1 a_2 x$ cannot be the empty string for any choice of x since $a_1 a_2$ is not empty. So there is no inverse of any non-empty string with respect to concatenation.

Where two binary operations are defined on a set S there may be a structure property involving both operations.

Definition 10.8 (Distributive property) *The operation $\square$ is said to be distributive over $\triangle$ on a set S if and only if*

$$x \square (y \triangle z) = (x \square y) \triangle (x \square z)$$

for all $x, y, z \in S$.

Example 1: Multiplication is distributive over addition on the real numbers.

$$\forall x, y, z \in \mathbf{R}, \quad x(y + z) = xy + xz$$

Example 2: Addition is not distributive over multiplication on the real numbers.

Counterexample: $2 + (3 \times 4) = 14$ but $(2+3) \times (2+4) = 30$.

Example 3: Intersection is distributive over union on the power set of A.

For $X, Y, Z \in \mathcal{P}(A)$

$$X \cap (Y \cup Z) = (X \cap Y) \cup (X \cap Z)$$

Example 4: Union is distributive over intersection on the power set of A.

For $X, Y, Z \in \mathcal{P}(A)$

$$X \cup (Y \cap Z) = (X \cup Y) \cap (X \cup Z)$$

Generalised properties

The distributive property and the associative property may each be generalised to apply to more than three elements.

Generalised distributive property Let the operation $\square$ be distributive over $\triangle$ on the set S. Then the general distributive property below is also true.

$$x\square(y_1\triangle \cdots \triangle y_n) = (x\square y_1)\triangle \cdots (x\square y_n)$$

for all $x, y_1, \cdots y_n \in S$.

The proof is by induction on n. The best way to do the proof is to set up the repeated sum as a recursion, but we will be satisfied to accept the result without proof.

Generalised associative property Let the binary operation $\square$ on S be associative. Then for each sequence $x_1, x_2 \cdots x_n \in S, (n \geq 3)$, with any arrangement of parentheses (satisfying correct parenthesis rules) the following expression has the same value:

$$x_1\square x_2\square \cdots \square x_n$$

The proof is by induction on n. The number of equal expressions for n elements is 2^{n-2}. For example, for four elements the result is

$$(a\square(b\square c))\square d = a\square(b\square(c\square d)) = ((a\square b)\square c)\square d = (a\square b)\square(c\square d)$$

Now that we have defined the basic 'structure properties' of operations on a set, and proved some associated results it is time to define an 'algebra' formally.

Definition 10.9 (Algebra) *An algebra is a set of objects S, one or more operations $(\Box, \triangle, \cdots)$ on S and some structure properties satisfied by these operations.*

A notation for an algebra is $(S, \Box, \triangle, \cdots)$. The possible structure properties are given in definitions 10.3−10.8.

10.1.2 Algebras with one operation

We look at three algebras that have one binary operation. Semigroups and monoids are used in formal languages and automata theory. Groups are used in automata and coding theory.

Definition 10.10 (Semigroup) *A semigroup $(S, \Box)$ is a set S together with an associative binary operation $\Box$ on S.*

Definition 10.11 (Monoid) *A monoid $(M, \Box)$ is a semigroup in which there is an identity for the operation $\Box$ on M.*

Definition 10.12 (Group) *A group $(G, \Box)$ is a monoid in which each element of the set has an inverse.*

If the operation is commutative then the group is said to be an Abelian or commutative group. The inverse of each element of a group is *unique* by theorem 10.2. The identity for a group is unique by theorem 10.1.

The word Abelian honours the Norwegian mathematician Niels Henrik Abel (1802-29) who used such groups in the theory of solving algebraic equations by means of the extraction of roots.

Example 1: $(A^+, \text{concatenation})$ is a semigroup.

1. Concatenation is associative; so the algebra is a semigroup.

2. A^+ consists of the set of all strings over the alphabet A but does not include the 'empty' string. Therefore there is no identity for this algebra and it is nothing more than a semigroup.

Example 2: $(A^*, \text{concatenation})$ is (a semigroup and) a monoid but not a group.

1. Concatenation is associative; so the algebra is a semigroup.

2. A^* consists of the set of all strings over the alphabet A including the 'empty' string, λ. We showed that the empty string λ is the identity for this algebra in example 4 (p. 227). Therefore the algebra is a monoid.

3. Not every element has an inverse. As a counterexample consider the string $\alpha = car$. There is no string x that could follow '*car*' to produce '', the empty string.

    Therefore the algebra is not a group.

Example 3: The integers form a group with addition, $(Z, +)$. The real numbers also form a group with addition, $(R, +)$. Each of these groups is Abelian.

Example 4: The integers without 0 form a group with multiplication. The notation for the group is $(Z - \{0\}, \times)$. The real numbers without 0 form a group with multiplication, $(R - \{0\}, \times)$.

Example 5: The set, P, of all permutations on a finite set

$$A = \{a_1, a_2, \cdots, a_n\}$$

with function composition, is a group.

1. Permutations were discussed in section 9.4.2 and a permutation was there defined to be a function. Function composition is associative (see example 1, p. 224). Therefore P is a semigroup under function composition.

2. The permutation

$$\iota = \begin{pmatrix} a_1 & a_2 & \cdots & a_n \\ a_1 & a_2 & \cdots & a_n \end{pmatrix}$$

    is an identity, so P is a monoid.

3. Each permutation has an inverse, because the permutation is a one-one function from A onto A. To see this consider a permutation σ. Suppose σ maps a_i to a_j. We write $\sigma : a_i \mapsto a_j$. Define $\sigma^{-1} : a_j \mapsto a_i$, and, because σ is one-one and onto, σ^{-1} is well defined. Clearly

$$\sigma \circ \sigma^{-1} = \sigma^{-1} \circ \sigma = \iota$$

    So σ^{-1} is the inverse of σ, and P is a group under function composition.

For $n > 2$ the permutation group is not Abelian. The number of elements in the group is finite (equal to $n!$) and so we say it is a *finite* group.

10.1.3 Groups

Groups are worth some deeper study. They have sufficient structure to be interesting and they are present as substructures in many more complex algebras. Rather than use definition 10.12, which refers back to the definitions of 'monoid' and 'semigroup', it is preferable to list all the properties of a group in one place.

Properties of the group(G,□)

1. □ is associative.

2. There exists an $e \in G$, called the identity with respect to □, such that

   $$e \square x = x = x \square e \quad \text{for all } x \in G.$$

3. For each $x \in G$ there exists a unique element $y \in G$ such that

   $$x \square y = e = y \square x.$$

   The element y is called the inverse of x with respect to □.

   If □ is a 'multiplication type' operation, the inverse of x is usually written x^{-1}.

   If □ is an 'addition type' operation, the inverse of x is usually written $(-x)$.

4. (Optional property) If $x \square y = y \square x$ for all $x, y \in G$, then the group is said to be Abelian, or commutative.

Example: Given the operation table for ×,

×	a	b
a	a	b
b	b	a

$(\{a, b\}, \times)$ is a group.

The operation is closed on the set.

The operation is associative:

$$a \times (a \times b) = a \times b = b \text{ and } (a \times a) \times b = a \times b = b$$

$$\text{Therefore} \quad a \times (a \times b) = (a \times a) \times b.$$

Eight proofs of this sort, to cover all sequences of a's and b's, are required to establish associativity for the group.

The identity is a.

Each element in the set has an inverse; $a^{-1} = a, \quad b^{-1} = b.$

As in ordinary arithmetic, repeated applications of an operation may be represented conveniently by the notation, p^m where p belongs to a group $(G, \square)$ and m is a non-negative integer.

Definition 10.13 (Exponentiation) *For $p \in (G, \square)$ a group with identity e, and an integer $m \geq 0$:*

$$p^m = \begin{cases} e & \text{if } m = 0 \\ p^{m-1} \square p & \text{if } m \geq 0 \end{cases}$$

If $\square$ is an 'addition type' operator then we usually write mp not p^m.

Example 1: $p^3 = p^2 \square p = p \square p \square p$.

The parentheses are omitted since $\square$ is associative.

Example 2: $p^1 = e \square p = p$.

Definition 10.14 (Negative exponentiation) *For $p \in (G, \square)$ a group, and integer $m < 0$:*
$$p^m = (p^{-m})^{-1}$$

The value of p^{-m} is given by definition 10.13. The notation $(\quad)^{-1}$ indicates 'the inverse of the element in parentheses with respect to $\square$ on G'.

Theorem 10.4 *If p is an element of the group $(G, \square)$ and m, n are integers then:*

1. $(p^m)^{-1} = (p^{-1})^m$

2. $p^m \square p^n = p^{m+n}$

3. $(p^m)^n = p^{mn}$

Proof: Each part of this theorem can be proved by the (generalised) associative property of $\square$ and definitions 10.13, 10.14, and (for part 1) theorem 10.3.

The results of the theorem are easy to remember since they are identical in appearance to the results for exponentiation on real numbers.

Theorem 10.5 *For p in the group $(G, \square)$:*

$$p^m \square p^n = p^n \square p^m \quad \text{for } m, n \in \mathbf{Z}.$$

Proof: From theorem 10.4, part 2, and the commutative property of addition on $\mathbf{Z}$:

$$p^m \square p^n = p^{m+n} = p^{n+m} = p^n \square p^m$$

for all integers m, n.

Theorem 10.6 *Let p be an element of a finite group $(G, \square)$ and let n, the number of elements in G, satisfy $n > 1$, then there exists an $s > 0$ such that $p^s = e$ where e is the identity for $\square$ on S and $s \leq n$.*

Proof: The sequence $p, p^2, \cdots p^{n+1}$ contains $n + 1$ elements each of which is in G (by closure).

G has n elements, so (by the pigeonhole principle) two of the elements in the sequence must be the same.

Suppose $p^a = p^b$ with $0 < a < b$.

Let $s = b - a \quad (> 0)$.

$$
\begin{aligned}
\text{Then } p^s = p^{b-a} &= p^b \square p^{-a} && \text{theorem 10.4 part 2}\\
&= p^b \square (p^a)^{-1} && \text{definition 10.14}\\
&= p^b \square (p^b)^{-1} && \text{argument above}\\
&= e && \text{definition 10.7}
\end{aligned}
$$

also $0 < a < b \leq n + 1$ so $s = b - a \leq n$ and the theorem is proved.

Note: If $n = 1$ in the above theorem then $s = 0$, but the result is not very interesting.

Theorem 10.7 (Cyclic group) *Let p be an element of a group $(S, \square)$, with at least two elements, and suppose s is the smallest positive integer such that $p^s = e$ where e is the identity for $\square$ on S. Then:*

$$C = \{e, p, p^2, \cdots p^{s-1}\}$$

with operation $\square$ is an Abelian group. $(C, \square)$ is called a cyclic *group and p is called a* generator *of the group.*

Proof: The existence of s is guaranteed by theorem 10.6.

1. The operation $\square$ on S is associative. Every element in C is also in S. Therefore $\square$ on C is associative.

2. There is an identity (e) in C.

3. Each element of C has an inverse.

For $0 \le m \le s$, $\quad p^m \square p^{s-m} = p^{m+s-m} = e = p^{s-m} \square p^m$,

by theorem 10.4 part 2 and definition 10.13,

so p^{s-m} is the inverse of p^m.

4. The group is Abelian by theorem 10.5.

Corollary: If S is finite and $|S| = n$ we say the group $(S, \square)$ is a finite group of order n. From theorems 10.5 and 10.6 every finite group is either a cyclic group itself or contains a cyclic group within it.

Example 1: $(\{e, a, b\}, +)$ with the operation table

+	e	a	b
e	e	a	b
a	a	b	e
b	b	e	a

is a cyclic group.

From the table, $a^2 = b$, and $a^3 = a^2 + a = b + a = e$
or using 'additive' notation, $3a = (a + a) + a = b + a = e$.
By theorem 10.6, $(\{e, a, a^2\}, +)$ is a cyclic group with generator a.

Equally, $b^3 = e$, so the same group may be written $(\{e, b, b^2\}, +)$ and b is a generator for the group.

Example 2: Define:

$$e = \begin{pmatrix} 1 & 2 & 3 \\ 1 & 2 & 3 \end{pmatrix}, \quad a = \begin{pmatrix} 1 & 2 & 3 \\ 2 & 3 & 1 \end{pmatrix}, \quad b = \begin{pmatrix} 1 & 2 & 3 \\ 3 & 1 & 2 \end{pmatrix}.$$

Then the permutation group $(\{e, a, b\}, \circ)$ has the identical operation table as the previous example. The connection between these groups is much stronger than the fact that the elements of the first can be mapped one-one onto the second. The groups are said to be *isomorphic* (defined below).

Example 3: $(\mathbf{Z}_m, +)$ is a cyclic group.

This is the general result corresponding to example 1.

Definition 10.15 (Isomorphism) *Two groups are isomorphic if and only if there is a one-one function mapping the elements of one onto the* *elements of the other with the property that the function maps the 'product' of any two elements in one group to the product of the images of those elements in the other group.*

If the groups are $(G_1, \square)$ and $(G_2, \triangle)$ and the function ϕ maps G_1 onto G_2 then, for all x, y in G_1:

$$\phi(x \square y) = \phi(x) \triangle \phi(y).$$

The function ϕ is called an isomorphism of the group G_1 onto the group G_2.

10.2 ORDERED STRUCTURES

In the previous sections we defined operation properties (commutativity, associativity, and so on) and identified different algebras (semigroups, monoids and groups) by the properties of the operation defined on them.

In this section we look at structures in which a *partial order relation* is important.

10.2.1 Partially ordered sets and lattices

Recall from Chapter 3 that a partial order relation (p. 58) is reflexive, antisymmetric and transitive. A set on which such a relation is defined is called a **partially ordered set**. (Some people contract this to *poset*.)

Definition 10.16 (Partially ordered set) *A partially ordered set, (S, ρ), is a set S on which is defined a partial order, ρ.*

The notation (S, ρ) is intended to be different to that for an algebra. The Greek letter ρ is used to indicate a partial order relation and should not be confused with an operation symbol. Some writers use the symbol $\preceq$ or the symbol $\subseteq$ in place of ρ.

Example 1: $S = \{1, 2, 3, 4, 5, 6, 7, 8, 9, 10\}$; ρ is the partial order relation 'divides', where 'divides' means 'divides exactly with no remainder'. For example $2\rho 2, 2\rho 6, 3\rho 6$ but not $4\rho 2$ and not $5\rho 7$.

The relation 'divides' is a partial order because:

1. $a\rho a$ for all a in S;

2. if $a\rho b$ and $b\rho a$ then $a = b$;

3. if $a\rho b$ and $b\rho c$ then $a\rho c$.

Divides is not a total order on this set because, for example, neither $5\rho 7$ nor $7\rho 5$.

Example 2: $S = \{1, 2, 3, 5, 6, 10, 15, 30\}$, ρ is the partial order 'divides'.

The relation ρ is not a total order on this set S because, for example, neither $2\rho3$ nor $3\rho2$.

Example 3: S = the power set of $\{a, b, c\}$, ρ = 'is a subset of' (ρ is the relation $\subseteq$).

The relation ρ is a partial order because:

1. $A \subseteq A$ for all A in S;

2. if $A \subseteq B$ and $B \subseteq A$ then $A = B$ in S;

3. if $A \subseteq B$ and $B \subseteq C$ then $A \subseteq C$ in S.

But ρ is not a total order because, for example:

$$\{a, b\} \not\subseteq \{a, c\} \text{ and } \{a, c\} \not\subseteq \{a, b\}$$

Now in none of the examples is the relation a total order. However each partial order does contain some sequences that are totally ordered. For example, the sequences $1, 2, 4, 8$ and $1, 2, 10$ in example 1 are totally ordered.

In those sequences, if we choose two elements from the sequence one element divides the other. If the elements are the same we know that every number divides itself. And if they are different the one on the left divides the one on the right.

In example 2: $1, 2, 6, 30$ and $1, 2, 10, 30$ are totally ordered sequences.

In example 3: $\emptyset, \{a\}, \{a, b\}, \{a, b, c\}$ and $\emptyset, \{a\}, \{a, c\}, \{a, b, c\}$ are totally ordered sequences.

The totally ordered sequences may be illustrated by a *Hasse* diagram (named after Helmut Hasse (1898−1979), a German mathematician). The Hasse diagram for a partially ordered set (S, ρ) is a graph in which the vertices represent the elements of S and two vertices a, b are joined by an edge if $a\rho b$; except that:

1. since it is known that ρ is reflexive it is not considered necessary to draw a loop corresponding to $a\rho a$; and

2. since it is known that ρ is transitive it is not considered necessary to draw an edge for $a\rho c$ if both $a\rho b$ and $b\rho c$ are included.

Traditionally the graph has been drawn without arrows and with the 'least' element at the 'bottom'. Here (Fig. 10.2) it has been drawn with the 'least' element at the left and 'increasing' to the right.

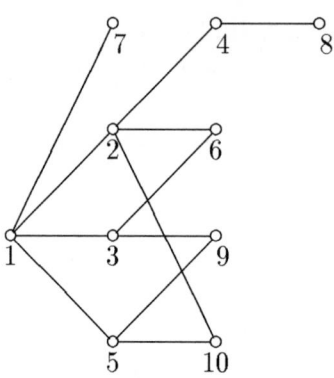

Figure 10.2: 'Hasse' diagram for example 1

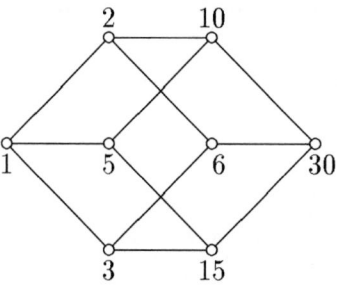

Figure 10.3: example 2

In example 1, 1 divides 2, 2 divides 4, 4 divides 8, so 1,2,4,8 may be regarded as an 'increasing' sequence with respect to the relation 'divides'. The corresponding path in the Hasse diagram (Fig. 10.2) goes (more or less) left to right.

Definition 10.17 (Greatest lower bound (glb)) *Let (S, ρ) be a partially ordered set. For any $a, b \in S$ an element $g \in S$ with the following properties is called a* greatest lower bound *for a, b.*

1. *$g\rho a$ and $g\rho b$*

2. *$\forall x \in S[(x\rho a) \wedge (x\rho b)] \rightarrow x\rho g$*

Example: From Figure 10.2 for example 1 it may be seen that
        glb(4,10) = 2; glb(2,5) = 1; glb(6,3) = 3; glb(10,10) = 10.

Theorem 10.8 *If (S, ρ) is a partially ordered set and $a, b \in S$ then (a, b) has at most one glb.*

Proof: Suppose $x, y \in S$ are glb's of a, b.

Then by definition 10.17   part 1: $x\rho a$, $x\rho b$ and $y\rho a$, $y\rho b$

part 2: $x\rho y$ and $y\rho x$

So by antisymmetry $x = y$.

Definition 10.18 (Least upper bound (lub)) *Let (S, ρ) be a partially ordered set. For any $a, b \in S$ an element $l \in S$ with the following properties is called a* least upper bound *for a, b.*

1. *$a\rho l$ and $b\rho l$*

2. *$\forall x \in S[(a\rho x) \wedge (b\rho x)] \rightarrow l\rho x$*

Example: From Figure 10.2 for example 1 it may be seen that
lub(4,10) is not defined ; lub(2,5) = 10; lub(6,3) = 6; lub(7,7) = 7.

Example: From Figure 10.3 for example 2 it may be seen that
lub(6,10) = 30; lub(2,5) = 10.

Theorem 10.9 *If (S, ρ) is a partially ordered set and $a, b \in S$ then (a, b) has at most one lub.*

Proof: Suppose $x, y \in S$ are lub's of a, b.

By definition 10.18 and an argument similar to that of theorem 10.8

$$x\rho y \text{ and } y\rho x$$

By antisymmetry $x = y$.

Both lub and glb map pairs of elements in S to unique elements of S. If lub and glb are defined for all pairs in S they can be regarded as binary operations on S. In this case let us have an infix operator symbol for them. Notation, if lub and glb are binary operations on S then:

$$\text{lub}(a, b) = a \triangleleft b$$

$$\text{glb}(a, b) = a \triangleright b$$

The structure for which every pair of elements has both a glb and an lub is called a lattice.

Definition 10.19 (Lattice) *A lattice is a partially ordered set (L, ρ), in which every pair of elements in L has a least upper bound and a greatest lower bound.*

A lattice (L, ρ) has two binary operations defined on it, and so we may construct from it the algebra $(L, \triangleleft, \triangleright)$.

Examples: 1. The partially ordered set of example 1 is not a lattice because not every pair of elements has a least upper bound. (For example lub(6,10) is not defined on the set.)

    2. The partially ordered set of example 2 with partial order relation 'divides'(ρ) is a lattice. The binary operations are 'greatest common divisor'($\triangleleft$) and 'least common multiple' ($\triangleright$).

    3. The partially ordered set of example 3 with relation 'is a subset of'(ρ) is a lattice. The binary operations are 'intersection' ($\triangleleft$) and 'union' ($\triangleright$).

The partially ordered set in example 1, (S, ρ) is not a lattice. As observed above, the pair $(6,10)$ has no upper bound in S. We could enlarge S to include 30, but then further enlargements would be necessary to include other upper bounds. If S were enlarged to the set of all integers $\mathbf{Z}$ then the structure would be a lattice. For every pair of integers has a least upper bound and a greatest lower bound in $\mathbf{Z}$, so $(\mathbf{Z}, \rho)$ is a lattice. But the set is unbounded above and below so there is no greatest element and no least element.

Definition 10.20 (Least, greatest element) *Let (S, ρ) be a partially ordered set. An element m is called the* greatest *(or maximum) element if for all $a \in S, a\rho m$. An element s is called the* least *(or minimum) element if for all $a \in S, s\rho a$.*

In example 1 the least element is 1 but there is no greatest element.

In example 2 the greatest element is 30 and the least element is 1.

In example 3 the greatest element is $\{a, b, c\}$ and the least element is $\emptyset$.

The modified Hasse diagram for example 3 is drawn in Figure 10.4 illustrating that lattice.

10.2.2 Boolean algebras

The structure that we consider next is called a Boolean algebra in honour of George Boole. Boole published *An Investigation of the Laws of Thought* in 1854, in which he set up a logical algebra of this type. Boolean algebras are used in information processing and switching theory.

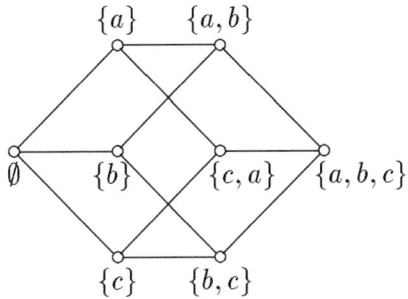

Figure 10.4: Example 3

A Boolean algebra is not *required* to have an order relation defined on it. It is included in section 10.2 because it has an interesting association with lattices.

Definition 10.21 (Boolean algebra) *A Boolean algebra is an algebra* $(B, +, \cdot, ^-, 0, 1)$ *where* $+$ *and* $\cdot$ *are binary operations on B and are commutative, associative and idempotent. Each binary operation is distributive over the other. The constant 0 is the identity for* $+$ *and 1 is the identity for* $\cdot$*. The operation* $^-$ *on B is a unary operation and satisfies the properties*

$$a + \overline{a} = 1 \text{ and } a \cdot \overline{a} = 0 \quad \forall a \in B.$$

The set B must contain at least the two elements 0, 1. The symbols may be considered as 'nullary' operations mapping from the empty set to 0,1 in B. In this way the notational scheme for an algebra is preserved. The following are examples of Boolean algebras.

Example 1: $(\{0, 1\}, +, \times, ^-, 0, 1)$ with the operation tables

+	0	1
0	0	1
1	1	0

×	0	1
0	0	0
1	0	1

$\overline{0} = 1, \overline{1} = 0$

Example 2: (power set of $A, \cup, \cap$, complement on sets, $\emptyset, A$); is a Boolean algebra.

The structure in example 3 on page 237 is a Boolean algebra of this type with $A = \{a, b, c\}$.

Example 3: $(S, +, \cdot, ^-, 1, 30)$; where $S = \{1, 2, 3, 5, 6, 10, 15, 30\}$; $a + b$ denotes 'the least common multiple of a, b'; $a \cdot b$ denotes 'the greatest common divisor of a, b' and $\overline{x}$ denotes the number $\frac{30}{x}$ is a Boolean algebra. (Refer to example 2, p. 237.)

The following theorems give general properties that hold for all Boolean algebras. They may be proved from definition 10.21.

Theorem 10.10 (Complement) *For all x in $(B,+,\cdot,^-,0,1)$, a Boolean algebra*

$$x'' = x.$$

Theorem 10.11 (Absorption) *For all x,y in $(B,+,\cdot,^-,0,1)$, a Boolean algebra*

$$x \cdot (x + y) = x, \quad x + (x \cdot y) = x$$

Theorem 10.12 (De Morgan) *For all x,y in $(B,+,\cdot,^-,0,1)$, a Boolean algebra*

$$\overline{(x \cdot y)} = \overline{x} + \overline{y}, \quad \overline{(x + y)} = \overline{x} \cdot \overline{y}$$

10.2.3 Boolean algebras and lattices

In this section we look at the construction of a Boolean algebra from a lattice and of a lattice from a Boolean algebra.

Starting from a lattice

It has been shown that for a lattice (L,ρ), glb($\lhd$) and lub($\rhd$) are binary operations on L. It may be proved that these operations on L are commutative, associative and idempotent.

If, as well, each operation is distributive over the other, the lattice is called a *distributive lattice*.

If the lattice has a greatest element (1) and a least element (0) then a unary operation, *complement($'$)* may be defined on L so that for each element $a \in L$ there exists an element a' in L, such that

$$a \lhd a' = 0 \text{ and } a \rhd a' = 1$$

(i.e. glb(a,a') = 0 and lub(a,a') = 1). A lattice with a least element, a greatest element and a complement operation on the set is called a *complemented* lattice.

Now if (L,ρ) is a distributive, complemented lattice with least element 0 and greatest element 1, it can be shown that $(L,\rhd,\lhd,',0,1)$ is a Boolean algebra.

Examples 2 and 3 from the set of Boolean algebras given on page 241 have been derived in this way from the lattices of examples 2 and 3 on page 237.

Starting from a Boolean algebra

Let $(S, +, \cdot, ^-, 0, 1)$ be a Boolean algebra.
For $a, b \in S$, define $a \preceq b$ if and only if $a \cdot b = a$.
The relation '$\preceq$' so defined is a partial order.

Proof: We must prove the propositions in the Boolean algebra that translate to the reflexive, antisymmetric and transitive properties of the relation under the definition above.

1. $a \cdot a = a$ ($\cdot$ is idempotent).
   So by the definition, $a \preceq a$ and '$\preceq$' is reflexive.

2. If $a \cdot b = a$ and $b \cdot a = b$
   then $a = b$ (since $\cdot$ is commutative).
   So by the definition, if $a \preceq b$ and $b \preceq a$ then $a = b$ and '$\preceq$' is antisymmetric.

3. If $a \cdot b = a$ and $b \cdot c = b$
$$
\begin{aligned}
a \cdot (b \cdot c) &= a \cdot b & \text{(replacing } b \cdot c \text{ by } b) \\
\text{then } (a \cdot b) \cdot c &= a \cdot b & \text{(associative)} \\
a \cdot c &= a & \text{(replacing } a \cdot b \text{ by } a)
\end{aligned}
$$

   So by the definition, if $a \preceq b$ and $b \preceq c$ then $a \preceq c$ and '$\preceq$' is transitive.

The above definition of '$\preceq$' has *induced* a partial order on S.

To prove that $(S, \preceq)$ is a lattice it is necessary to show that for all $a, b \in S$ both $\text{lub}(a, b)$ and $\text{glb}(a, b)$ exist. The next theorem asserts that they do exist and provides an expression for each.

Theorem 10.13 (glb, lub) *Let $(S, \preceq)$ be the partially ordered set induced on the Boolean algebra $(S, +, \cdot, ^-, 0, 1)$ by the relation '$\preceq$' defined by $a \preceq b$ if and only if $a \cdot b = a$.*
Then $\forall a, b \in S$, glb and lub exist, defined by

    1. $glb(a, b) = a \cdot b$

    2. $lub(a, b) = a + b$

and $(S, \preceq)$ is a lattice.

Proof: 1.(glb) By the associative, commutative and idempotent properties of '$\cdot$'

$$(a \cdot b) \cdot a = a \cdot (b \cdot a) = a \cdot (a \cdot b) = (a \cdot a) \cdot b = (a \cdot b)$$

and

$$(a \cdot b) \cdot b = a \cdot (b \cdot b) = (a \cdot b)$$

By definition of '$\preceq$'

$$(a \cdot b) \preceq a \text{ and } (a \cdot b) \preceq b$$

So the first part of definition 10.17 is satisfied.
Now suppose that $x \preceq a$ and $x \preceq b$.
By definition of '$\preceq$', $x \cdot a = x$ and $x \cdot b = x$ so $x \cdot a = x \cdot b$.
We wish to show $x \preceq a \cdot b$.
Now $x \cdot (a \cdot b) = (x \cdot a) \cdot b = (x \cdot b) \cdot b = x \cdot (b \cdot b) = x \cdot b = x$
and $x \preceq a \cdot b$.
So the second part of definition 10.17 is satisfied, and

$$a \cdot b = \text{ glb}(a, b)$$

2. (lub). The proof is set as an exercise. It is similar to part 1 but uses $+$ and definition 10.18.

It follows immediately that $(S, \preceq)$ is a lattice.

Theorem 10.14 *The lattice $(S, \preceq)$ has least element 0 and greatest element 1.*

Proof: (1) The least element is 0.
By a complement property and the commutative, associative and idempotent properties of '$\cdot$', for all $a \in S$,

$$0 \cdot a = (a \cdot \overline{a}) \cdot a = a \cdot (a \cdot \overline{a}) = (a \cdot a) \cdot \overline{a} = a \cdot \overline{a} = 0$$

so by definition, $\quad 0 \preceq a \quad \forall a \in S$, and 0 is the least element.
    (2) The greatest element is 1.
This proof is given as an exercise.

Two further propositions about $(S, \preceq)$ may be proved but are left as exercises.

1. $(S, \preceq)$ is a distributive lattice.

2. $(S, \preceq)$ is a complemented lattice.

10.3 FINITE FIELDS

A field is an algebra in which the four binary operations addition, multiplication, subtraction and division are defined on the set. Another way to describe the field structure is to define it as a group $(S, +)$ together with a group $(S - \{0\}, \times)$ where the binary operation $\times$ is distributive over $+$.

Definition 10.22 (Field) *If $(S, +)$ is a group and $(S - \{0\}, \times)$ is a group and $\times$ is distributive over $+$ then the algebra $(S, +, \times)$ is called a field.*

The obvious examples of a field are the number systems **R** and **Q**; (but not **Z** since $(\mathbf{Z} - \{0\}, \times)$ is not a group). We know that each of **R** and **Q** has a (total) order relation, '$\leq$', defined on it, but there is nothing in the definition of a field that will give rise to this property. Some additional properties must be imposed on a field if it is to have an order relation. The following is one way to do it.

Definition 10.23 (Ordered field) *A field with set F is said to be ordered if it contains a non-empty set P (the positives) which is closed with respect to the addition and multiplication operations of the field and $P, \{0\}, F - P$ is a partition of F in which each cell is non-empty.*

The partial order relation '$\leq$' is then defined by

$$\forall x, y \in F, \quad x \leq y \text{ if and only if } y - x \in P \cup \{0\}$$

While it is possible to derive all the familiar properties of the order relation '$\leq$' from the definition, that wouldn't give us anything new. The purpose here is merely to show that a field is not necessarily ordered. The following are examples of finite fields. None of the examples is ordered.

Example 1: $S = \{0, 1\}$ and define addition and multiplication by the tables:

+	0	1		$\times$	0	1
0	0	1		0	0	0
1	1	0		1	0	1

Then $(S, +, \times)$ is a field.

Example 2: $S = \{0, 1, 2\}$ and define addition and multiplication by the tables:

+	0	1	2		$\times$	0	1	2
0	0	1	2		0	0	0	0
1	1	2	0		1	0	1	2
2	2	0	1		2	0	2	1

Then $(S, +, \times)$ is a field.

The field in example 2 is not ordered. Noting that $1 + 2 = 0$ we observe that the inverse of 1 with respect to addition is 2, i.e. $(-1) = 2$. We might try to fit the definition of an ordered field by defining $P = \{1\}$. Then $\{1\}, \{0\}, \{2\}$ is a partition of S. But P is not closed under addition, for example $1 + 1 = 2$.

Example 3: The set of residues congruent modulo m, where m is prime, with addition and multiplication modulo m is a field.

10.4 EXERCISES

1. Show that division on the set of real numbers is not associative.

2. Give an example to illustrate the statement '*concatenate* is associative on A^* '.

3. What is meant by the statement 'Intersection is idempotent on the power set of A'?

4. Let $L = \{1, 3, 5, 7, 15, 21, 35, 105\}$ and let ρ be the relation 'divides' on L. Then (L, ρ) is a partially ordered set. Draw a Hasse diagram for (L, ρ). Find lub(3,7), lub(15,21), glb(7,21). Find the least and the greatest elements of the set. Find a complement operation for the lattice.

5. Let $\mathbf{B} = \{\mathsf{T}, \mathsf{F}\}$.
   Consider the algebra $(\mathbf{B}, \wedge, \vee)$.

   (a) State in symbols what is meant by each of the following.
       i. $\wedge$ is commutative.
       ii. $\vee$ is associative.
       iii. $\vee$ is distributive over $\wedge$.
   (b) Does $\wedge$ have an identity? If so, what?
   (c) Does $\vee$ have an identity? If so, what?
   (d) Give a reason why $(\mathbf{B}, \vee)$ is not a group.

6. Addition on $M_{2\times 2}$, the set of 2 by 2 matrices with real elements, is defined by

$$\begin{pmatrix} a_{11} & a_{12} \\ a_{21} & a_{22} \end{pmatrix} + \begin{pmatrix} b_{11} & b_{12} \\ b_{21} & b_{22} \end{pmatrix} = \begin{pmatrix} a_{11} + b_{11} & a_{12} + b_{12} \\ a_{21} + b_{21} & a_{22} + b_{22} \end{pmatrix}$$

   Consider the algebra $(M_{2\times 2}, +)$.

   (a) What is the identity for addition?
   (b) Does each matrix in $(M_{2\times 2}+)$ have an inverse?
   (c) Is addition commutative?
   (d) Is addition associative?
   (e) Is addition idempotent?
   (f) Is $(M_{2\times 2}, +)$ a semigroup? a monoid? a group? an Abelian group?

7. What structure properties hold for the function 'append' regarded as an operation on lists?

8. Here is a 'multiplication' table for an algebra $(A, *)$ with $A = \{a, b, c\}$.

$*$	a	b	c
a	a	a	b
b	a	b	c
c	b	c	b

(a) Is $*$ commutative?

(b) Is $*$ associative?

(c) Is $*$ idempotent?

(d) Does $*$ have an identity?

(e) Does each $x \in A$ have a unique inverse?

9. Given that the operation $\square$ on S is associative and $a, b, c, d, e \in S$ write down all the expressions with five terms equal to

$$(a \square (b \square (c \square (d \square (e \square))))).$$

10. Use the fact that multiplication on real numbers is associative to prove that for $a, b, c, d, e, f \in \mathbf{R}$

$$(a \times b) \times ((c \times d) \times (e \times f)) = a \times (b \times (c \times (d \times (e \times f))))$$

11. Prove that a cyclic group is Abelian.

12. In the proof of theorem 10.4, part 1, we must show that

$$p^m \square (p^{-1})^m = e = (p^{-1})^m \square p^m$$

A proof may be constructed by assuming $m \geq 0$ for case 1 and $m < 0$ for case 2. Case 1 is done by the generalised associative property. A proof of case 2 uses definition 10.14, theorem 10.3 and definition 10.7. Set out the proof of case 2.

13. Given a set A, $(\mathcal{P}(A), \subseteq)$ is a lattice. What is meant by the statement $\cup$ on $\mathcal{P}(A)$ is associative? Is the statement true?

14. Let (L, ρ) be a lattice. Prove:

(a) $\mathrm{lub}(\triangleright)$ on L is commutative.

(b) glb(◁) on L is idempotent.

15. Let S be the set $\{1, 2, 3, 5, 6, 10, 15, 30\}$. Let $a + b$ denote 'the least common multiple of a, b'; $a \cdot b$ denote 'the greatest common divisor of a, b' and $\overline{x}$ denote 'the number $\frac{30}{x}$'. Then $(S, +, \cdot, ^-, 1, 30)$ is a Boolean algebra. (Example 3 on page 241.)

Prove the following:

   (a) The identity for $+$ is 1.

   (b) The identity for $\cdot$ is 30.

   (c) The operation $+$ is idempotent.

   (d) $\forall x \in S \quad x + \overline{x} = 30$.

   (e) $x \cdot \overline{x} = 1$.

16. Prove that for all x in a Boolean algebra $(B, +, \cdot, ^-, 0, 1)$

   (a) $x.0 = 0$ (b) $x + 1 = 1$

17. Theorem 10.11 (Absorption) states:

For all x, y, z in a Boolean algebra $(B, +, \cdot, ^-, 0, 1)$

$$(a) \quad x \cdot (x + y) = x, \quad (b) \quad x + (x \cdot y) = x$$

Prove part (a).

18. Theorem 10.12 (de Morgan) states:

For all x, y, z in a Boolean algebra $(B, +, \cdot, ^-, 0, 1)$

$$(a) \quad \overline{(x \cdot y)} = \overline{x} + \overline{y}, \quad (b) \quad \overline{(x + y)} = \overline{x} \cdot \overline{y}$$

Prove part (a).

19. Let $(S, +, \cdot, ^-, 0, 1)$ be a Boolean algebra.
Define the relation '$\succeq$' on S by

$$a \succeq b \text{ if and only if } a + b = a$$

Prove that the relation so defined induces a partial order on S.

20. Let $(S, \preceq)$ be the lattice induced on the Boolean algebra $(S, +, \cdot, ^-, 0, 1)$ by the relation '$\preceq$' defined by

$$a \preceq b \text{ if and only if } a \cdot b = a.$$

Prove:

   (a) glb$(a, b) = a + b$ for all $a, b \in S$.

   (b) 1 is the greatest element of $(S, \preceq)$.

Chapter 11

Reasoning and Correctness

In 1854 George Boole published *An Investigation of the Laws of Thought*. The opening paragraph is:

> The design of the following treatise is to investigate the fundamental laws of those operations of the mind by which reasoning is performed; to give expression to them in the symbolical language of a Calculus, and upon this foundation to establish the science of Logic and construct its method; to make that method itself the basis of a general method for the application of the mathematical doctrine of Probabilities; and, finally, to collect from the various elements of truth brought to view in the course of these inquiries some probable intimations concerning the nature and constitution of the human mind.

Boole had more in mind than an algebra of 'true and false' for which he is now chiefly remembered. Many logicians since Boole have contributed to the development of symbolic logic as presented here.

The development of 'artificial intelligence' and the production of fault-free software that may be maintained and modified economically are important current endeavours in computing. The first has resulted in a close examination of human reasoning. The second has led to the development of formal methods for specification and related methods for guaranteeing the correctness of software. These are active areas of research, and opinions differ about what part mathematics may play in solving the problems. However, if mathematics is to make a contribution, the most promising area to explore is that of logic and proof.

The aim in this chapter is to introduce mathematics related to human reasoning, and program verification so that you will have some understanding of the issues and some knowledge of tools being developed in these areas.

We look first at an inference scheme intended to copy ordinary reasoning.

11.1 NATURAL DEDUCTION

This section describes a plan for proving propositions correct, different from that given in Chapter 4. In the earlier approach a number of logical laws were taken as axioms (i.e. were assumed to be true) and results were obtained using the inference rules for 'replacement' and 'transitivity'. Here only inference rules are used with no axioms or assumed logical formulas. In fact they are consistent approaches producing the same results.

We may think of an inference scheme as an *operational scheme* for deducing or inferring new propositions. The propositions involved may be any logical expression (usually involving connectives) as introduced in Chapter 4. Then the inference rules may be interpreted as instructions on how to operate correctly on the propositions to produce new propositions.

Introduction (I)	*Elimination* (E)
$\wedge$-I	$\wedge$-E
$$\frac{p, q}{p \wedge q}$$	$$\frac{p \wedge q}{p, q}$$
$\vee$-I	$\vee$-E
	$[p]\quad[q]$
$$\frac{p}{p \vee q}$$	$$\frac{p \vee q,\ r,\quad r}{r}$$
$\rightarrow$-I	$\rightarrow$-E
$[p]$	
$$\frac{q}{p \rightarrow q}$$	$$\frac{p, p \rightarrow q}{q}$$
$\equiv$-I	$\equiv$-E
$$\frac{p \rightarrow q, q \rightarrow p}{p \equiv q}$$	$$\frac{p \equiv q}{p \rightarrow q, q \rightarrow p}$$
$\neg$-I	$\neg$-E
$[p]$	$[\neg p]$
$$\frac{q \wedge \neg q}{\neg p}$$	$$\frac{q \wedge \neg q}{p}$$

Table 11.1: Rules for natural deduction

Gerhard Gentzen formulated the rules in 1934-35. They consist of rules for the *introduction* of connectives and rules for the *elimination* of connectives. Each rule consists of one or more premises separated by a line from a single conclusion or *inference*. A premise in brackets [ ] indicates an assumption.

The ∧-introduction rule states that if a proposition p is proved and a proposition q is proved then we may infer $p \wedge q$. The ∧-elimination rule states that if $p \wedge q$ is proved then we may infer p and we may also infer q. There are two ∨-introduction rules. The first states 'from p infer $p \vee q$' and the second states 'from q infer $p \vee q$'. Both rules are necessary. The ∨-elimination rule models 'proof by cases' and is often used in mathematics. The rule states that if $p \vee q$ is proved, and it is proved that r follows from p and that r follows from q, then we may infer r.

The →-introduction rule states that if q follows from the assumption of p then we may infer $p \rightarrow q$. This rule is used in the inductive step in mathematical induction. The →-elimination rule is called *modus ponens*. It states that if p and $p \rightarrow q$ are both proved then we may infer q. The ¬-introduction and ¬-elimination rules are called 'proof by contradiction' and are often used in mathematical proof where the existence or non-existence of some fact is to be proved. The method is used in the proof that $\sqrt{2}$ is irrational and in the proof that there are more primes than any natural number. The ≡-introduction rule states that if $p \rightarrow q$ and $q \rightarrow p$ are both proved then we may infer that $p \equiv q$. A converse is the rule for ≡-elimination which states that if $p \equiv q$ is proved then we may infer $p \rightarrow q$ and $q \rightarrow p$.

In the proof examples below each line begins with a number for reference. Then a proposition is asserted. At the end of the line a reason for the assertion is given. The reason is either a premise (or assumption) or a list of the reference numbers of previous propositions on which the assertion depends together with the inference rule used.

Example 1: Prove $p \wedge q \rightarrow p \vee q$.

The first line of the proof is an *assumption* because it is planned to use the rule →-I to introduce the conditional (→).

Proof:
1. $p \wedge q$ .... assumption
2. p .... 1, ∧-E
3. $p \vee q$ .... 2, ∨-I
4. $p \wedge q \rightarrow p \vee q$ .... 1,3, →-I

Assertion 3 follows from the assumption in line 1 allowing →-I to be invoked on the next line.

The following example illustrates a more complex use of assumptions in order to use ∨-E and ¬-E. Those two rules are employed because the premises contain an '∨' and a '¬' and the conclusion (q) contains no connective at all.

Example 2: Prove that from $p \lor q, \neg p$ we may infer q.

Proof:
1. $p \lor q$ premise 1
2. $\neg p$ premise 2
3. p 1st assumption for $\lor$-E
4. $\neg q$ assumption for $\neg$-E
5. $p \land \neg p$ 2,3, $\land$-I
6. q 4,5, $\neg$-E
7. q 2nd assumption for $\lor$-E
8. q 7
9. q 1,3,6,7,8, $\lor$-E

Indentation is used to draw attention to an assumption in the proof. When the consequence of that assumption is realised the proof returns to the previous alignment. The assumption is then said to be *discharged*. When the rule $\lor$-E is invoked (line 9) the consequences of two assumptions (lines 3, 7) are realised. On line 6 the assumption on line 4 is discharged. All assumptions have led to a conclusion and the argument returns to the original alignment. In a mathematical proof line 3 and line 7 would be labelled case 1 and case 2 respectively.

Example 3: Prove $p \equiv p$.   (The law of identity.)

Proof:
1. p assumption for $\rightarrow$-I
2. p 1
3. $p \rightarrow p$ 1,2, $\rightarrow$-I
4. $p \rightarrow p$ 3
5. $p \equiv p$ 3,4, $\equiv$-I

Example 3 proves the law of identity from section 4.2. At first sight the proof may seem unduly repetitious, for example lines 3,4. However, to prove $p \equiv q$, both the 'forward' implication ($p \rightarrow q$) and the 'backward' implication ($q \rightarrow p$) are required so two propositions are needed in the proof. We could remove line 4 provided the reason for line 5 read '3,3, $\equiv$-I'.

Example 4: Prove $p \lor \neg p$.   (The law of the excluded middle.)

Proof:
1. $\neg(p \lor \neg p)$ assumption for $\neg$-E
2. $\neg p$ assumption for $\neg$-E
3. $p \lor \neg p$ 2, $\lor$-I
4. $(p \lor \neg p) \land \neg(p \lor \neg p)$ 1,3, $\land$-I
5. p 2,4, $\neg$-E
6. $p \lor \neg p$ 5, $\lor$-I
7. $(p \lor \neg p) \land \neg(p \lor \neg p)$ 1,6, $\land$-I
8. $p \lor \neg p$ 1,7, $\neg$-E

The law of the excluded middle is another axiom given in section 4.2.

Example 5: Prove $\neg p \vee q \equiv p \to q$. (The law of implication.)

The proof has been paragraphed into subproofs labelled (a), (b), (c) to simplify the reading.

Proof: (a) Prove $(\neg p \vee q) \to (p \to q)$.

1.	$\neg p \vee q$	assumption for $\to$-I
2.	p	assumption for $\to$-I
3.	q	1,2, example 2
4.	$p \to q$	2,3, $\to$-I
5.	$(\neg p \vee q) \to (p \to q)$	1,4, $\to$-I

(b) Prove $(p \to q) \to (\neg p \vee q)$.

6.	$p \to q$	assumption for $\to$-I
7.	$p \vee \neg p$	example 4
8.	$\neg p$	1st assumption for $\vee$-E
9.	$(\neg p \vee q)$	8, $\vee$-I
10.	p	2nd assumption for $\vee$-E
11.	q	6,10, $\to$-E
12.	$\neg p \vee q$	11, $\vee$-I
13.	$\neg p \vee q$	7,8,9,10,12, $\vee$-E
14.	$(p \to q) \to (\neg p \vee q)$	6,13, $\to$-I

(c) Prove $\neg p \vee q \equiv p \to q$.

15.	$\neg p \vee q \equiv p \to q$	5,14, $\equiv$-I

Example 6: Prove $p \to (q \wedge r) \equiv (p \to q) \wedge (p \to r)$.

The following proof could be paragraphed: (a) lines 1-9, (b) lines 10-17, (c) line 19.

Proof:

1.	$p \to q \wedge r$	assumption for $\to$-I
2.	p	assumption for $\to$-I
3.	$q \wedge r$	1, $\to$-E
4.	q	3, $\wedge$-E
5.	r	3, $\wedge$-E
6.	$p \to q$	2,4, $\to$-I
7.	$p \to r$	2,5, $\to$-I
8.	$(p \to q) \wedge (p \to r)$	6,7, $\wedge$-I
9.	$(p \to (q \wedge r)) \to (p \to q) \wedge (p \to r)$	1,8, $\to$-I
10.	$(p \to q) \wedge (p \to r)$	assumption for $\to$-I
11.	$(p \to q)$	10, $\wedge$-E
12.	$(p \to r)$	10, $\wedge$-E

$$
\begin{array}{lll}
13. & p & \text{assumption for } \rightarrow\text{-I} \\
14. & q & 13,11, \rightarrow\text{-E} \\
15. & r & 13,12, \rightarrow\text{-E} \\
16. & q \wedge r & 14,15, \wedge\text{-I} \\
17. & p \rightarrow (q \wedge r) & 13,16, \rightarrow\text{-I} \\
18. & (p \rightarrow q) \wedge (p \rightarrow r) \rightarrow (p \rightarrow (q \wedge r)) & 10,17, \rightarrow\text{-I} \\
19. & p \rightarrow (q \wedge r) \equiv (p \rightarrow q) \wedge (p \rightarrow r) & 9,18, \equiv\text{-I}
\end{array}
$$

The rules for natural deduction given so far involve only *statements* or *propositions*. This appears to exclude quantified predicates, a very common form of mathematical proposition. However, in mathematics we can turn predicates into propositions by treating each variable as an *arbitrary* constant. The description 'arbitrary' applied to a constant means that it satisfies no special conditions. The argument may then be pursued using the rules of natural deduction to infer a conclusion for the arbitrary variables. It is argued that the proposition is true for all possible values of the variable and the appropriate quantifier may be introduced. This provides a rule for $\forall$-I.

Table 11.2 gives introduction and elimination rules for the quantifiers introduced in Chapter 4 and is included to show how the natural deduction approach may be extended to deal with quantified predicates; although a full treatment will not be given.

Introduction (I)	*Elimination* (E)
$\forall$-I	$\forall$-E
$\dfrac{p(x) \; (x \text{ arbitrary})}{\forall x \, p(x)}$	$\dfrac{\forall x \, p(x)}{p(a)}$
$\exists$-I	$\exists$-E
$\dfrac{p(a)}{\exists x \, p(x)}$	$\dfrac{p(x) \rightarrow c, \; (x \text{ arbitrary})}{\exists x \, p(x)}$
	c
	(if x is not free in c)

Table 11.2: Quantifier rules

In Table 11.2 the predicate $P(x)$ contains a variable x and possibly other variables.

In $\forall$-*elimination* and $\exists$-*introduction*, a denotes an *allowable* substitution for x in $P(x)$. Full rules for describing allowable substitutions are complex. Briefly, we must avoid clashes between different occurrences of the same variable. Such clashes may be introduced by the substitution. For example, 'x is the largest element of the set S' can be expressed in predicate calculus by the formula '$P(x) = \forall a \in S\,(a \leq x)$'. Substituting a for x yields '$\forall a \in S\,(a \leq a)$', which does not yield the intended meaning for $P(a)$. We first need to change the variable a in $P(x)$, to some other, say b. Then $P(x) = \forall b \in S\,(b \leq x)$ and $P(a)$ becomes '$\forall b \in S\,(b \leq a)$'.

In $\forall$-*introduction* and $\exists$-*elimination*, the statement 'x arbitrary' means that x has not occurred 'free' in any of the assumptions used to make the given inferences. A definition of 'free' suitable for automatic operation, is complicated. Intuitively the assumptions must not restrict x.

In $\exists$-*elimination* it is also required that x be not 'free' in c. The rule $\forall$-I is called the rule of *generalisation* and $\forall$-E is called *instantiation*. The example below shows two of the quantifier rules in use.

Example 7: Prove: $\forall x \in \mathbf{R}\ x^2 \geq 0$.

The order axioms and a theorem for real numbers are taken as 'premises' (i.e. given facts) for the proof. The premises hold for all real numbers a, b, c. The argument of the proof is to show the result holds for $x \geq 0$ and for $x \leq 0$, and hence for all x.

Premises:

1	$a \geq 0$ or $a \leq 0$	total order axiom
2	if $a \geq b$ and $c \geq 0$ then $ac \geq bc$	order axiom
3	if $a \leq b$ and $c \leq 0$ then $ac \geq bc$	order axiom
4	$0.a = 0$	theorem

Proof:

1.	$\forall x(x \geq 0 \vee x \leq 0)$	premise 1
2.	$x \geq 0 \vee x \leq 0$	1, $\forall$-E
3.	$x \geq 0$	1st assumption for $\vee$-E
4.	$x^2 \geq 0.x$	3, premise 2
5.	$x^2 \geq 0$	4, premise 4
6.	$x \leq 0$	2nd assumption for $\vee$-E
7.	$x^2 \geq 0.x$	6, premise 3
8.	$x^2 \geq 0$	7, premise 4
9.	$x^2 \geq 0$	2,3,5,6,8, $\vee$-E
10.	$\forall x\ x^2 \geq 0$	9, $\forall$-I (x arbitrary)

Remarks: It would not be valid to use $\forall$-I at line 5 since x is not arbitrary here owing to the assumption at line 3. Similarly, it would not be valid to use $\forall$-I at line 8 since x is not arbitrary here owing to the assumption at line 6. In line 9 the assumptions have been discharged and it is valid to apply $\forall$-I to obtain line 10. Line 10 completes the proof.

In addition to reflecting a 'natural' method of reasoning, the system described in this section is closer to the way in which mathematicians argue.

Mathematical reasoning itself has a long history, beginning 500-300 BC in Ancient Greece.

The propositions below have been chosen to illustrate mathematical argument. The emphasis is on the *rules* of reasoning employed.

Example 8: Prove: *If a, x, b belong to a group $(G, \square)$ and*

$$a \square x = b$$

then $x = a^{-1} \square b$.

Premises: 1 $(G, \square)$ is a group.
            2   $a \in G$
            3   $b \in G$
            4   $x \in G$
            5   $a \square x = b$

Premise 1 contains a number of simpler facts. It could be enlarged to give the properties of a group, namely $\square$ is a binary operator on G; $\square$ is associative; there is an identity e for $\square$ in G and for each x in G there is a unique inverse in G. The properties could be defined further by stating the definitions of words such as 'associative', 'inverse' and 'identity'.

Proof: 1. $x \in G$.   (premise 4)
         2. $\square$ is the operator for G.   (premise 1)
         3. e is the identity for $\square$ in G.   (premise 1)
         4. $x = e \square x$.   (lines 1,2,3 and definition 10.6)
         5. $a \in G$.   (premise 2)
         6. $e = a^{-1} \square a$.   (definition 10.7)
         7. $x = (a^{-1} \square a) \square x$.   (lines 4,6, rule of replacement)
         8. $x = a^{-1} \square (a \square x)$.   (line 7, premise 1, associativity)
         9. $x = a^{-1} \square b$ (line 8, premise 3,5, rule of replacement)

Conclusion: $x = a^{-1} \square b$.

Remarks: The formal proof above is a sequence of propositions each supported by a statement of the premises and earlier propositions that establish its truth. Many mathematicians would prefer to take as 'understood' the statement of premises as the 'if part' of the proposition, together with an implicit sequence of earlier axioms, definitions and theorems of the area of mathematics under discussion. They would then be satisfied with lines 4,7,8,9 as an indication of the proof. Although correct, the other lines seem to clutter the proof with trivia. The proof containing every detail, including the enlargement of premise 1, is called a *formal* proof and at present it is the formal proof that is required for computer 'reasoning' of this type. Programs using this level of detail have been used to prove mathematical theorems and to discover new theorems.

In the following examples of proof we will change our focus from the fine detail of individual steps to the logical structure of 'rules of reasoning'.

Example 9 (Proof by cases):

Theorem 11.1 *If a is an element of a group $(G, \square)$ and m is an integer, then*

$$(a^m)^{-1} = (a^{-1})^m$$

Proof: Consider an arbitrary integer m.

Case 1: $m \geq 0$

The generalised associative law and repeated applications of $a \square a^{-1} = e$ and $a \square e = a$ allow us to deduce that

$$a^m \square (a^{-1})^m = e$$

and hence $\quad (a^m)^{-1} = (a^{-1})^m$

Case 2: $m < 0$

Definitions 10.13, 10.14, case 1 and theorem 10.3 allow us to deduce

$$(a^m)^{-1} = (a^{-1})^m$$

Case 1 and case 2 cover all possible values for m.

Therefore: $(a^m)^{-1} = (a^{-1})^m$ for an arbitrary integer m.

(It follows that the proposition is true for all integers m, since the proof used no properties specific to any particular m.)

Remarks: The logical structure of this proof follows the scheme:

$$\frac{p \vee q, \qquad \begin{array}{cc} [p] & [q] \\ r & r \end{array}}{\text{Therefore:} \quad r}$$

where, for an arbitrary integer m,

p is the statement $(m \geq 0)$
q is the statement $m < 0$
r is the statement $(a^m)^{-1} = (a^{-1})^m$
$[x]$ denotes 'assuming x holds true'

and the scheme may be read 'if p or q is true and given p we can deduce r and given q we can deduce r, then r is in fact true'.

The argument eliminates the 'or' connective so *or-elimination* has been employed. The comment in parentheses at the end of the proof asserts that the rule of generalisation may now be applied.

Example 10: The 'inductive step' in a proof by induction contains the following sequence.

Assume $P(k)$ is true for a constant $k \geq 0$.

$$\vdots \quad \left.\right\} \text{ a sequence of deductions}$$

leading to the conclusion $\quad P(k+1) \quad$ is true.

Therefore $P(k) \to P(k+1)$ for any constant $k \geq 0$.

Therefore $P(n) \to P(n+1)$ for all $n \geq 0$.

Remarks: The logical structure of this proof follows the scheme,

$$\frac{\begin{array}{c}[p]\\ q\end{array}}{\text{Therefore:} \quad p \to q}$$

where, for an arbitrary integer k:

p is the statement $P(k)$
q is the statement $P(k+1)$
$[x]$ denotes 'assuming x holds true'

and the scheme may be read 'if when p is assumed to be true, we can deduce q, then it follows that $p \to q$ is true'.

The argument introduces the 'if-then' connective taking k to be constant, so the $\to$-*introduction* rule has been employed.

It is valid to employ the rule $\forall$-I in the conclusion since k is arbitrary after the inductive assumption $(P(k))$ has been discharged.

Example 11 (Proof by contradiction):

Theorem 11.2 (Theorem 10.1) *If there is an identity 'e' for an operator $\square$ over a set S then that identity is unique.*

Proof: Suppose that e, f are identities for $\square$ in S and suppose $e \neq f$. Then:

$$
\begin{aligned}
e &= e \square f &&\text{since } f \text{ is an identity} \\
&= f &&\text{since } e \text{ is an identity}
\end{aligned}
$$

But $e \neq f$. Contradiction! Therefore if an identity exists it is unique.

Remarks: The logical structure of this proof follows the scheme:

$$
\frac{\begin{array}{c}[\neg p] \\ q \wedge \neg q\end{array}}{\text{Therefore:} \quad p}
$$

where, p is the statement 'e, f are identities for $\square$ over S'
  q is the statement '$e \neq f$'
and the scheme may be read 'assume p is false (i.e. $\neg p$ true); then deduce a contradiction, for example $q \wedge \neg q$. It follows that p is true'.

The argument eliminates the 'negation' connective so therefore employs the *$\neg$-elimination* rule.

Example 12: Let $(B, +, ., ^-, 0, 1)$ be a Boolean algebra. Then for all x in B, $0.x = 0$.

Proof:
1.	$0.x = x.0$	. is commutative
2.	$= x.(x.\overline{x})$	complement property
3.	$= (x.x).\overline{x}$	. is associative
4.	$= x.\overline{x}$	. is idempotent
5.	$= 0$	complement property

Each line of the proof relies on *modus ponens*. In general this takes the form:

$$
\frac{\begin{array}{c}x, \overline{x}, 0 \text{ belong to a Boolean algebra} \\ \textbf{if } a, b, c \text{ belong to a Boolean algebra } \textbf{then} \text{ they satisfy} \\ \text{the properties of a Boolean algebra, including the} \\ \text{distributive and commutative properties of . and } +.\end{array}}{\begin{array}{c}\text{Therefore (for example):} \\ x.(x.\overline{x}) = (x.x).\overline{x} \quad \text{(associative property).}\end{array}}
$$

Example 13: Define the predicate $P(n), n \in \mathbf{N}$ by

$$
P(n) \equiv (\underbrace{1 + 3 + 5 + \cdots + (2n - 1)}_{n \text{ terms}} = n^2)
$$

Prove $\forall n \geq 1\ P(n)$.

Proof: 1. $P(1)$ $(1 = 1^2)$ is true
   2. $P(k)$ $(k$ fixed) assumption for $\to$-I
   3. $(1 + 3 + \cdots + (2k - 1) = k^2)$ 2
   4. $(1 + 3 + \cdots + (2k - 1) + (2k + 1))$
         $= k^2 + 2k + 1$ 3
         $= (k + 1)^2$ algebra
      i.e. $P(k + 1)$ is inferred
   5. $P(k) \to P(k + 1)$ 2,4, $\to$-I
   6. $\forall k \geq 1(P(k) \to P(k + 1))$ 5, $\forall$-I, (generalisation)
   7. $\forall n \geq 1 P(n)$ 1,6, PMI

where PMI stands for the principle of mathematical induction.

11.2   CORRECTNESS OF ALGORITHMS

Proof of correctness for algorithms (and programs) has developed its own style. Some writers (e.g. Backhouse (1986) and Gries (1981)) argue that better software is produced if it is developed with the proof in mind. The main problem with the proof methods that have been developed so far is that they are cumbersome and hard to apply to a program of reasonable size.

We have hitherto regarded an algorithm as a sequence of instructions that solves a problem in a finite time. The value of an algorithm tends to be proportional to its generality, so we required an algorithm to accept input for each variable from a set of possible values. Then the *correctness* of an algorithm can only be decided relative to some statement of what it is supposed to do for each set of possible values. This statement is called a *specification*. The specification does not usually say anything about how the result is obtained but does say how the output must relate to the input and what inputs are to be accepted.

To avoid input/output problems we will adopt the simplification that an algorithm is a sequence of instructions to be applied to a set of variables of various types. Again for simplification we will consider only assignment instructions (including evaluation of expressions), branch (if ... then ... else) and loop (while) instructions. At any instant each variable has a value which is called its *state* at that time. The *state* of an algorithm refers to the set of values of each variable associated with the algorithm. The *specification* is now regarded as the relation between the initial state and the final state of an algorithm (i.e. before and after execution of the algorithm) together with a description of those initial states that lead to successful execution of the algorithm. The relation between initial state and final state is described using a *postcondition*. Valid initial states are identified by a *precondition* on the variables. As before, a precondition and a postcondition are each predicates, which together form the specification of an algorithm.

If for each execution of an algorithm (or program) that starts with the same initial state there is a unique final state, the algorithm is said to be *deterministic*. Otherwise, if the outcome may vary, the algorithm is said to be *non-deterministic*. A deterministic algorithm evaluates a function and in this section we consider only deterministic algorithms.

{P} A {Q}

The notation

$$\{P\}A\{Q\}$$

means that if the variables of algorithm (or program) A initially satisfy the precondition P, and the sequence of one or more instructions (A) is executed, then the algorithm will terminate with the postcondition Q satisfied.

A precondition is a predicate on the initial values and a postcondition is a predicate on both initial and final values of the variables of the algorithm. The expression $\{P\}A\{Q\}$ is itself a predicate.

Example 1:

$$\{a = a_0\}$$

$$a \leftarrow a * a; \ a \leftarrow a * a; \ a \leftarrow a * a$$

$$\{a = a_0^8\}$$

In this example the precondition is $P \equiv (a = a_0)$ and the postcondition is $Q \equiv (a = a_0^8)$. The 'variable' a_0 does not appear in the algorithm and is not changed by the algorithm. It serves to relate the final value of a to the initial value of a. The variable a_0 is a constant for a given execution of the algorithm.

Example 2:

$$\{(r = a = a_0 \geq 0) \wedge (b = b_0 > 0) \wedge (q = 0) \wedge (b_0 \leq a_0 < 2b_0)\}$$

$$q \leftarrow q + 1; \ r \leftarrow r - b$$

$$\{(a_0 = b_0 q + r) \wedge (0 \leq r < b_0)\}$$

The precondition is

$$P \equiv (r = a = a_0 \geq 0) \wedge (b = b_0 > 0) \wedge (q = 0) \wedge (b_0 \leq a_0 < 2b_0)$$

and the postcondition is

$$Q \equiv (a_0 = b_0 q + r) \wedge (0 \leq r < b_0).$$

The variables a_0, b_0 are not changed by the algorithm. After execution $q = 1$ and $r = a_0 - b_0$ and these values satisfy the postcondition.

The inequality $b_0 < a_0 < 2b_0$ is 'stronger' than the inequality given in the precondition in the sense that it excludes a possible value for a_0 (namely $a_0 = b_0$). The inequality $b_0 \leq a_0 < 2b_0$ given in the precondition is *weakest* in the sense that it allows all possible initial values that will cause the algorithm to terminate successfully with values that satisfy the postcondition.

Many preconditions may go with A and Q. However, we would like a precondition that allows every set of input values that will lead to termination of A with values satisfying Q.

Definition 11.1 (Weakest precondition) *If Q is a predicate and A an instruction (or sequence of instructions) then* the weakest precondition of A with respect to Q *is a predicate that is true for all initial values that guarantee that A will terminate satisfying Q. The weakest precondition of A with respect to Q is written $wp(A, Q)$.*

Example 3: Find $wp(A, Q)$ for each of the following, given that all variables are integer.

    (a) $A : a \leftarrow a + 1,\quad Q \equiv (a > n)$

    (b) $A : a \leftarrow a + 1; b \leftarrow b - 1,\quad Q \equiv a + b = 0$

    (c) $A : a \leftarrow a + 1; b \leftarrow b - 1,\quad Q \equiv a \times b = 0$

Solution:

    (a) The instruction increases the value of a by 1. So if $(a > n)$ is true initially it will remain true after execution of A. In addition, if $a = n$ initially, execution of A will result in $a > n$. Therefore:

$$wp(A, Q) \equiv (a \geq n) \equiv P.$$

    (b) The algorithm increases a by 1 and decreases b by 1, so $a + b$ is unchanged by execution of A. The expression $a + b$ is said to be *invariant with respect to A*. Therefore:

$$wp(A, Q) \equiv (a + b = 0) \equiv Q \equiv P.$$

(c) If $a = a_0, b = b_0$ initially then after execution $a = a_0+1, b = b_0-1$ and Q asserts that $(a_0 + 1) \times (b_0 - 1) = 0$. So initially $a_0 = -1$ or $b_0 = 1$. Therefore the precondition is:

$$wp(A, Q) \equiv (a_0 = -1) \vee (b_0 = 1) \equiv P.$$

Assignment instruction

Example 3 suggests a method for calculating the weakest precondition for an assignment instruction. If x and e are the same type of variable and if the expression e is well defined then:

$$wp(a \leftarrow e, Q(a)) \equiv Q(e) \quad \text{and}$$

$$\{Q(e)\}\, x \leftarrow e\, \{Q(x)\}.$$

'The same type of variable' means, for example, 'both are integer' or 'both are Boolean'. The requirement that e be well defined excludes values of the variables in the expression for e for which e does not exist, for example due to 'division by zero' or 'taking the square root of a negative'.

Example 4: Solve example 3 using $wp(a \leftarrow e, Q(a)) = Q(e)$.

(a) $wp(a \leftarrow a + 1, (a > n)) \equiv (a + 1 > n) \equiv (a \geq n)$.

(b) $wp(a \leftarrow a + 1; b \leftarrow b - 1, (a + b = 0)) \equiv ((a + 1) + (b - 1) = 0)$
$\equiv (a + b = 0)$.

(c) As argued above the weakest precondition is $((a+1)(b-1) = 0)$.

Example 5: $wp(a \leftarrow 0, (a = 0)) \equiv (0 = 0) \equiv \mathsf{T}$.

In words this means that for every initial value of a, after execution of $a \leftarrow 0$ the postcondition is satisfied. We might say that the precondition T allows all values of the input variables.

Example 6: $wp(a \leftarrow 0, (a = 1)) \equiv (0 = 1) \equiv \mathsf{F}$.

Here no input value can produce an output to satisfy the postcondition.

Logical properties of weakest preconditions

Although there is no difficulty with the construction of $wp(A, Q)$ where A is an assignment, other instructions (e.g. loops) may be difficult, perhaps impossible. There are some general logical properties of $wp(A, Q)$ that sometimes help.

1. If $wp(A, Q) \equiv \mathsf{T}$ then every initial state leads to termination of A satisfying Q.

2. If $wp(A, Q) \equiv \mathsf{F}$ then no initial state leads to termination of A satisfying Q.

3. $\{wp(A, Q)\}A\{Q\}$ merely says in another way that the weakest precondition for A *is* a precondition for A.

4. From $\{P\}A\{Q\}$ it may be inferred that $P \rightarrow wp(A, Q)$. This inference is another way to say that every state that leads to termination of A satisfying Q is included by $wp(A, Q)$.

5. $wp(A, \mathsf{F}) \equiv \mathsf{F}$.

   The postcondition F means that there are no pairs in the relation between initial values and final values of the variables of A.

The following inference rules for weakest preconditions are given for reference.

1. $wp(A, Q \wedge R) \equiv wp(A, Q) \wedge wp(A, R)$.

2. $wp(A, Q \vee R) \equiv wp(A, Q) \vee wp(A, R)$.

   This result holds for deterministic algorithms.

3. If $Q \rightarrow R$ then $wp(A, Q) \rightarrow wp(A, R)$.

4. $wp(A, \neg Q) \equiv \neg wp(A, Q)$.

Sequential composition

Although this rule has been used informally above we should write it down for reference.

Suppose that the algorithm A consists of sequential instructions $A_1; A_2$. Then:

$$
\begin{aligned}
wp(A, Q) &\equiv wp(A_1; A_2, Q) \\
&\equiv wp(A_1, wp(A_2, Q))
\end{aligned}
$$

Example 7: Construct $wp(i \leftarrow i + 1; s \leftarrow s + 2i - 1, s = i^2)$.

$$
\begin{aligned}
wp(A, Q) &\equiv wp(i \leftarrow i + 1, s + 2i - 1 = i^2) \\
&\equiv s + 2(i + 1) - 1 = (i + 1)^2 \\
&\equiv s = (i + 1)^2) - 2(i + 1) + 1 \\
&\equiv s = (i + 1 - 1)^2) \\
&\equiv s = i^2
\end{aligned}
$$

Branch instruction

The instruction if B then A_1 else A_2 may be read ' if B then execute A_1 and if not B then execute A_2' and is called a branch. If the postcondition is Q then we have:

$$wp(\text{if } B \text{ then } A_1 \text{ else } A_2, Q) \equiv (B \rightarrow wp(A_1, Q)) \wedge (\neg B \rightarrow wp(A_2, Q))$$

Another argument leads to:

$$wp(\text{if } B \text{ then } A_1 \text{ else } A_2, Q) \equiv (B \wedge wp(A_1, Q)) \vee (\neg B \wedge wp(A_2, Q))$$

and the two formulas may be proved equivalent (see exercise 11.3.9).

Example 8: Find a weakest precondition for the branch instruction (A) below relative to the postcondition

$$Q \equiv (max \geq a) \wedge (max \geq b)$$

$$\text{if } a > b \text{ then } max \leftarrow a$$
$$\text{else } max \leftarrow b$$

$$
\begin{aligned}
wp(A, Q) & \\
\equiv \ & ((a > b) \rightarrow (a \geq a) \wedge (a \geq b)) \wedge ((a \leq b) \rightarrow (b \geq a) \wedge (b \geq b)) \\
\equiv \ & ((a > b) \rightarrow (a \geq b)) \wedge ((a \leq b) \rightarrow (a \leq b)) \\
\equiv \ & \top
\end{aligned}
$$

Loop instruction

A loop repeats the same sequence of instructions until a desired goal is achieved. For such a strategy to pay off two characteristics are needed. Some feature of the problem should benefit from repetitions of the same sequence of actions, implying some constant property of the problem that invites the same sequence of instructions. Some other changes produced by the loop must change the state of the system in such a way that a goal is reached and the loop then terminates.

In example 11.1.13 we proved

$$\forall_{n \geq 1}, (1 + 3 + \cdots + (2n - 1) = n^2)$$

That example suggests the following algorithm and a postcondition for the summation of the first n odd numbers.

algorithm W: while $i \neq n$
$$\begin{cases} i \leftarrow i + 1 \\ s \leftarrow s + 2i - 1 \end{cases}$$
$$\{Q \equiv (s = n^2)\}$$

Let us call the 'while' instruction W and the 'body of the loop' A. The body of the loop consists of the assignments $i \leftarrow i + 1$ (increase i by 1 each pass through the loop) and $s \leftarrow s + 2i - 1$ (add the next odd number to the current value of the sum s).

Let k be the remaining number of passes through A before termination of W; let P_k be the precondition at this point.

If $k = 0$ then $i = n$ and Q must hold.

$$P_0 \equiv ((i = n) \wedge (s = n^2))$$

If $k = 1$ then $i \neq n$ and the postcondition for the pass through W is P_0.

$$\begin{aligned} P_1 &\equiv& ((i \neq n) \wedge wp(A, P_0)) \\ &\equiv& ((i \neq n) \wedge (i + 1 = n) \wedge (s + 2(i+1) - 1 = n^2)) \\ &\equiv& ((i = n - 1) \wedge (s + 2(i+1) - 1 = (i+1)^2)) \\ &\equiv& ((i = n - 1) \wedge (s = (i+1)^2 - 2(i+1) + 1)) \\ &\equiv& ((i = n - 1) \wedge (s = i^2)) \end{aligned}$$

If $k = 2$ then $i \neq n$ and P_1 is the postcondition for the next pass through W.

$$\begin{aligned} P_2 &\equiv& ((i \neq n) \wedge wp(A, P_1)) \\ &\equiv& ((i \neq n) \wedge (i + 1 = n - 1) \wedge (s + 2(i+1) - 1 = (i+1)^2)) \\ &\equiv& ((i = n - 2) \wedge (s = i^2)) \end{aligned}$$

In general, for k passes remaining:

$$P_k \equiv ((i = n - k) \wedge (s = i^2))$$

Now P_k is the condition that guarantees Q after exactly k executions of A. Therefore:

$$\begin{aligned} wp(W, s = n^2) \\ \equiv \exists_{k \geq 0}, (i = n - k \wedge s = n^2) \end{aligned}$$

This is equivalent to $i \leq n$ and $s = i^2$. The predicate $(s = i^2)$ is true at the end of each execution of the body of the loop. It is called an *invariant* of the while loop. The invariant may be read as 'the sum of the first i odd numbers is i^2'. The predicate $(i \leq n)$ is the condition for termination of the loop and since i is increased on each pass through the loop it must become false after a finite number of steps and termination is guaranteed.

Example 9: Algorithm 1.1(p. 3) (*find the quotient and remainder when integer a is divided by integer b, where $a > 0, b > 0$*), is written out below. Let us find the weakest precondition and prove that the algorithm is correct.

    input: Integers a, b.
                Precondition P $\equiv$ ($a > 0$ and $b > 0$).
    output: Quotient q and remainder r.
                Integers q, r .
                Postcondition Q $\equiv (a = bq + r$ and $0 \leq r < |b|)$.

    method: 1. $q \leftarrow 0$
                2. $r \leftarrow a$
                3. while $r \geq b$
                4. $q \leftarrow q + 1$
                5. $r \leftarrow r - b$

If a k can be found such that

$$P_k \equiv wp(3; 4; 5, Q) \text{ and } P \rightarrow wp(1; 2, P_k)$$

then the algorithm is proved correct with respect to P, Q.

Let there be k executions of 3; 4; 5 to termination of the while loop and let P_k be the precondition at this point.

If $k = 0$ then $r < b$ and Q must hold since no further instructions are to be carried out.

$$
\begin{aligned}
P_0 &\equiv ((r < b) \wedge (a = bq + r) \wedge (0 \leq r < |b|)) \\
&\equiv (0 \leq r < b) \wedge (a = bq + r)
\end{aligned}
$$

If $k = 1$ then $r \geq b$ and the postcondition is P_0.

$$
\begin{aligned}
P_1 &\equiv (r \geq b) \wedge wp(4; 5, (0 \leq r < b) \wedge (a = bq + r)) \\
&\equiv (r \geq b) \wedge wp(q \leftarrow q + 1; r \leftarrow r - b, (0 \leq r < b) \wedge (a = bq + r)) \\
&\equiv (r \geq b) \wedge (0 \leq r - b < b) \wedge (a = b(q + 1) + (r - b)) \\
&\equiv (b \leq r < 2b) \wedge (a = bq + r)
\end{aligned}
$$

Similarly:

$$P_2 \equiv (2b \leq r < 3b) \land (a = bq + r)$$

$$\text{and in general } P_k \equiv (kb \leq r < (k+1)b) \land (a = bq + r)$$

$$\text{Therefore, } wp(3;4;5,Q) \equiv \exists_{k \geq 0}, (kb \leq r < (k+1)b) \land (a = bq + r)$$

$$\text{Also} \quad wp(1;2,P_k) \equiv \exists_{k \geq 0}, (kb \leq a < (k+1)b) \land (a = a)$$

$$\equiv (a \geq 0) \land (b > 0)$$

Now $P \to (a \geq 0) \land (b > 0)$ so the algorithm is correct. Further we have identified that the given precondition is not the 'weakest' so we can allow another initial state (namely $a = 0, b > 0$). The invariant for the loop is the predicate $I \equiv (a = bq + r)$.

11.3   EXERCISES

1. The proof below has been constructed using the Gentzen rules of natural deduction. Supply reasons for each line of the proof.

   Proposition: From $p \land q, p \to r$ infer $q \land r$.

   Proof: 1. $p \land q$
          2. $p \to r$
          3. p
          4. r
          5. q
          6. $q \land r$

2. Supply reasons for each line of the proof below.

   Proposition: From $p \land q, p \to r$ infer $r \lor (q \to r)$.

   Proof: 1. $p \land q$
          2. p
          3. $p \to r$
          4. r
          5. $r \lor (q \to r)$

3. Each of the following assertions uses one rule of inference. Supply a reason for each.

   (a) From $p \land q \land (c \lor d)$ infer $c \lor d$.

   (b) From $p \lor q, q \lor r, p \lor q \to c$ infer c.

   (c) From p, q, r infer $p \lor r$.

   (d) From the assumption $\neg(a \lor b)$ it follows that $p \land \neg p$. Infer $a \lor b$.

4. Prove that $\neg(p \vee q) \to \neg p \wedge \neg q$ using the Gentzen rules of inference.

5. Use the inference rules to prove:

(a) $p \wedge (p \to q) \to q$

(b) from $\neg q$ infer $q \to p$

6. Let $(B, +, ., ^-, 0, 1)$ be a Boolean algebra. Then for all x in B, $x+1 = 1$. Prove the proposition. What laws of inference are used?

7. Find $wp(A, Q)$ for each of the following.

(a) $A : a \leftarrow a - 1,$ $Q \equiv (a \leq 0)$

(b) $A : a \leftarrow a + 2; b \leftarrow b - 2,$ $Q \equiv a + b = 0$

(c) $A : a \leftarrow a + 2; b \leftarrow b - 2,$ $Q \equiv ab = 0$

(d) $A : z \leftarrow zj; i \leftarrow i - 1,$ $Q \equiv (zj^i = c)$

(e) $A :$ if n is even then $n \leftarrow n/2; y \leftarrow y \times y,$ $Q \equiv (y^n = c).$

8. Prove $wp(A, Q \wedge R) \equiv wp(A, Q) \wedge wp(A, R).$

9. Prove $(b \to p) \wedge (\neg b \to q) \equiv (b \wedge p) \vee (\neg b \wedge q).$

10. Prove $\{P\}A\{Q\}$ for the algorithm below.

$\{P \equiv (s = i * (i + 1)/2)\}$ (precondition)

Algorithm A: $i \leftarrow i + 1$
                   $s \leftarrow s + i$

$\{Q \equiv (s = i * (i + 1)/2)\}$ (postcondition)

11. The following algorithm is designed to calculate the product of two non-negative integers, m, n by repeated addition.

input: two positive integers m, n.
output: a positive integer p equal to the product mn.

$p \leftarrow 0$

$i \leftarrow n$

while $i > 0$ do
$$\begin{cases} i & \leftarrow & i - 1 \\ p & \leftarrow & p + m \end{cases}$$

The postcondition is $Q \equiv (p = mn)$. A suggested precondition is $P = (m \geq 0) \wedge (n \geq 0)$. Find an invariant for the loop. Show that the algorithm is correct with respect to P, Q. Is P a weakest precondition for the algorithm?

12. The following algorithm is designed to calculate $n!$ for integer $n \geq 0$.

    integers i, n, fac.

    while $i < n$ do
    $$\begin{cases} i & \leftarrow & i+1 \\ fac & \leftarrow & fac * i \end{cases}$$

    The postcondition is $Q \equiv (fac = n!)$. A suggested precondition is $P \equiv (i = 0) \wedge (fac = 1)$. Find an invariant for the loop. Show that the algorithm is correct with respect to P, Q and that the P given is a weakest precondition for the algorithm.

Appendix A

Writing Algorithms

An algorithm for a given problem consists of a sequence of instructions to solve the problem. For a programmer, an algorithm stands midway between a problem to be solved and the program needed to solve it. For a mathematician, an algorithm represents a constructive solution to a problem as opposed to an argument that a solution exists. Despite the importance of algorithms there is no agreed method for writing them. Perhaps the best guiding principle is that algorithms should be written in good, clear English. They should be readily understandable by others and by oneself.

An algorithm should be reasonably general, providing a solution for many problems. Therefore it should involve *variables* whose values can be reset to match each problem. The *instructions* should come from an agreed set of instructions, but writing an algorithm to solve a problem is a creative task and instruction sets tend to change to suit different problems and different writers.

The instructions are applied to constants and variables of stated *types*. The *type* of a variable is more than the set of possible values for the variable; it is also an instruction to use the correct operation for that type when an operation is invoked. This is important because we often use the same operation symbol to mean different things. For example, the operation 'addition of integers' is different to the operation 'addition of matrices' and to 'addition of real numbers'. But each is denoted by the same symbol, namely '+'. The meaning must be determined by the type of the variables on which the operation is imposed.

In mathematics the type of variable is often indicated informally by the choice of letter. For example, i, j, k, l, m, n are often integer variables; x, y are real and z is complex. However, it is safer to declare the type of each variable at the beginning of an algorithm.

Variables are represented by letters such as $a, b, c, \ldots$ or $x, y, z, \ldots$ and words such as *factor, prime, trialfactor, average, ...* In textbooks a variable is normally represented by a letter or word in italics. The use of words to represent variables helps the reader know what is intended in an algorithm or program and is popular in computing. But whether the variable behaves as the word used to represent it, depends on the instructions for its evaluation.

Instructions are ordinary English words together with some mathematical and other symbols. The set of instructions for an algorithm is usually informal and may be added to for individual problems if that helps communication. An instruction followed by another instruction is an instruction. Therefore any sequence of instructions is an instruction; and conversely it is possible (and often desirable) to *refine* an instruction into a sequence of (simpler) instructions.

The *assignment*, the *branch* and the *loop* are three fundamental instructions used almost universally in algorithms and these will be described in detail.

The *assignment* instruction assigns a value to a variable. The value may be given as a constant or the value of an expression.

Example 1: x, integer

$$\text{give } x \text{ the value } 0$$

In this example the variable x has been assigned the value 0. A special symbol, $\leftarrow$, has been set aside to convey this meaning. Thus example 1 may be written

$$x, \text{ integer}$$
$$x \leftarrow 0$$

Example 2:   *prime*, Boolean
      a, integer
      1. *prime* $\leftarrow$ F (i.e. false)
      2. $a \leftarrow 3$
      3. $a \leftarrow a \times a$
      4. $a \leftarrow a \times a$

Each of the numbered lines is an assignment and they are executed in that order. After line 1 the Boolean variable *prime* has the value 'false'. After line 2 has been executed the integer variable a has the value 3. After line 3 has been executed the new value of a is 9. After line 4, a has the value 81. In each case the 'current' value of a is used by the active instruction. While instruction lines are labelled here for easy reference, it is not necessary to number the lines in an algorithm in general.

A *branch* allows a dynamic change in the sequence of instructions during execution of an algorithm. The instruction 'if B then A_1 else A_2' is defined for a Boolean B and pair of instructions A_1, A_2. At execution the Boolean must have one of the values 'true' or 'false'. If B is 'true' then instruction A_1 is executed and A_2 is ignored. If B is 'false' then instruction A_2 is executed and A_1 is ignored. The instruction A_2 may be 'empty', that is may result in no action.

Example 3: x, real

                a, real

                1. if $x \geq 0$

                2.   then $x \leftarrow x - 2$

                3.   else $x \leftarrow x + 2$

                4. *(after the branch is complete execute line 4.)*

In line 1 the Boolean variable $(x \geq 0)$ is evaluated. If for the current value of x the Boolean is true, that is if x is not negative, then line 2 is executed, and line 3 is ignored. Lines 2 and 3 are indented to draw attention to the branch structure. If for the current value of x the Boolean is false, that is if x is negative, then line 3 is executed.

In the execution of line 2 it may happen that the value of x changes to 'negative'. This has no effect on the sequence of instructions followed. The sequence of instructions is only affected by the value of x at the time line 1 is executed.

A *loop* allows a sequence of instructions to be repeated.

Example 4: n, integer

                a, integer

                1. while $a \leq \sqrt{n}$

                2. $n \leftarrow n/2$

                3. $a \leftarrow a + 1$

                4. *(after the loop is complete, execute line 4)*

The sequence of lines 1,2,3 is called a 'while loop'. The sequence 2,3 is called the 'body of the loop'. In line 1 the Boolean variable $(a \leq \sqrt{n})$ is evaluated. If for the current values of a, n the Boolean is false, the execution of the loop is complete and the algorithm continues with line 4. Return to the original alignment offers a visual cue that the 'while' is completed. If for the values of the variables at line 1 the Boolean is true, then line 2 is executed followed by line 3. All the lines in the body of the loop are executed even if the Boolean becomes false at some point during this process. After the body of the loop has been executed the algorithm next executes line 1 in the same manner as described above.

Two common variations to the 'while' loop are 'repeat A until B' and 'for (*a given sequence of values of* i) do A_i'.

In the 'repeat' version the body (A) is always executed the first time. After completion of A the Boolean B is evaluated. If the Boolean is 'true' the loop is complete and the algorithm moves to the next instruction in the sequence. If the Boolean is 'false' then the body of the loop is executed again and the process continues as described.

In the 'for' version the body of the loop A_i is executed for each value of the variable i in the given sequence. It is considered bad practice to change the values of the variable i from within the body of the loop.

Appendix B

Answers to Exercises

Questions that ask for a calculation of some sort have a definite answer, and most exercises of this type have an answer supplied below.

Questions that ask for a proof or for an algorithm, have many correct answers and should be assessed for clarity and effective communication as well as for correctness. Sample proofs and sample algorithms have been given for some of the questions of this type. These answers should be correct but may always be improved for style, by which we usually mean effective communication. The most satisfactory way to check your proof is to see if it convinces a mathematically knowledgeable person. An algorithm should be tested by asking another person to use it.

Some questions are intended for investigation. Such questions may be assessed by a presentation or brief report. No answers have been provided for questions of this type.

Exercise Set 1.4 (page 13)

1. (a) $q = 2, r = 4$. 18, 7 are integers and $7 \neq 0$ so the precondition is T. $18 = 7 * 2 + 4$ and $0 \leq 4 < 7$ so the postcondition is T.

   (b) $q = -2, r = 4$; $-7 \neq 0$ precondition is T. $18 = (-7) * (-2) + 4$ and $0 \leq 4 < |-7|$ postcondition is T.

   (c) $q = -3, r = 3$; $7 \neq 0$ precondition is T. $-18 = 7 * (-3) + 3$ and $0 \leq 3 < 7$ postcondition is T

2. $\lfloor \sqrt{1987} \rfloor = 44$.

3. $11729 = 37 * 317$, no.

4. *For discussion.* The following is one of many possible answers:

input: Integer n.   (n is the number to be tested)
           A precondition is ($n > 1$).
output: A statement 'n is perfect' or 'n is not perfect'
method: Set a to 2.   (a is a possible factor)
           Set s to 1.   (s is the sum of factors found so far)
           While $a \leq \dfrac{n}{2}$:
                if a divides n exactly
                    then add a to s
                increase a by 1
           If $s = n$
                then report (n 'is perfect')
                else report (n ' is not perfect').

24 is not perfect. 28 is perfect.

5. $37, \lfloor 19 \cdot 9286 \rfloor = 19$.

6. 61 is prime. The algorithm passes through the loop 6 times (branch 2).
   $64 = 2^6$. The algorithm passes through the loop 5 times (branch 1).
   The result suggests that the algorithm may take more time to process a prime than a number with many factors.

7. *For discussion.*

input: Two integers in $\mathbf{Z}_7, a, b$.
           (a,b belong to $\{0,1,2,...6\}$)
output: s.   (s is the value of $a + b$ in $\mathbf{Z}_7$)
method: Let s be $a + b$.
           (s is to be the sum of a and b)
           If $s > 7 - 1$
                then replace s by s - 7

8. Greatest common divisor.

input: Integers a, b.
           A precondition is ($a > 0$ and $b > 0$).
output: g, the greatest common divisor of a, b.
method: While $a \neq b$:
                if $a > b$
                    then replace a by $a - b$
                    else replace b by $b - a$
           Give g the value a.

Exercise Set 2.4 (page 37)

1. (a) $1 * 10^3 + 2 * 10^2 + 0 * 10^1 + 9 * 10^0$

   (b) $1 * 2^5 + 1 * 2^4 + 0 * 2^3 + 1 * 2^2 + 0 * 2^1 + 1 * 2^0$

   (c) $12 * 16^4 + 1 * 16^3 + 0 * 16^2 + 10 * 16^1 + 10 * 16^0$

2. (a) $234_{10} = 11101010_2 = \overbrace{1110}\,\overbrace{1010} = EA_{16}$

   (b) $1024_{10} = 10000000000_2 = 400_{16}$

   (c) $255 = 256 - 1 = \overbrace{1111}\,\overbrace{1111}_2 = FF_{16}$

   (d) $747_8 = 111100111_2 = 1\,\overbrace{1110}\,\overbrace{0111}_2 = 1E7_{16}$

   (e) $326_8 = 11\,\overbrace{010}\,\overbrace{110} = \overbrace{1101}\,\overbrace{0110} = D6_{16}$

   Convert each digit to a binary word of length 3. Regroup in words of length 4 from the point. Convert each binary word of length 4 to a hexadecimal digit.

3. (a) 43981   (b) 5179   (c) 93

4. (a) 11111_2 (b) 1111_2 (c) 100100_2

   (d) quotient 10111_2, remainder 1.

5. (a) $BBBB_{16}$ (b) $1F8E_{16}$

6. 3020 (base 8)

7. 100 (base 2)

8. 63 is represented by 00111111. -17 is represented by 11101111.

9. 00111001 represents 57. 10011010 represents -102.

10. (a) If $a \neq 0, x = b/a$; if $a = 0$ and $b \neq 0$ then 'no solution'; if $a = 0$ and $b = 0$ then x may be any real number.

11. The notation 1; 24, 51, 10 means $1 + \dfrac{24}{60} + \dfrac{51}{60^2} + \dfrac{10}{60^3}$

$$\text{Babylonian value} = 1.414212962\cdots$$
$$\sqrt{2} = 1.414213562\cdots$$

The absolute error is 0.000001.

12. $215491/216169 = 1907/1913$

13. $y = 1110.0010012_2$. Relative error $= 0.004\%$.

14. 0.04%.

15. $\sqrt{5} = 2.236068\cdots$. $x_2 = 2.2361111$, and the absolute error is 0.00004.

16. $\dfrac{1351}{780}$. Error is 0.00000047.

17. (a) $914.0\dot{C}$ (base 16) (b) 34.9375 (base 10) (c) 110101010.001 (base 2)

18. (a) $0.\dot{1}00\dot{1}_2$ (b) The algorithm converts a decimal fraction to a binary fraction. If a repeats an earlier value then the output will cycle.

19.

$(n \neq 0)$	q	r	p	n	x
			1	13	3
T	6	1	3	6	3^2
T	3	0		3	3^4
T	1	1	3^5	1	3^8
T	0	1	3^{13}	0	3^{16}
F					

20. (a) $a = 2,\ 1.33333,\ 1.42857,\ 1.41176;\ y = 1.41176$.

(b) Absolute error is 0.002.

(c) To stop the algorithm. A rough measure of accuracy.

Exercise Set 3.5 (page 63)

1. (a) $\{m, a, t, h, e, i, c, l\}$ (b) $\emptyset, \{b\}, \{a\}, \{a, b\}$

(c) $\{n \mid n \text{ is a non-negative, integer power of } 2\}$

2. A is the set of integer multiples of 8. B is the set of integer multiples of 6. C is the set of integer multiples of 48. No; $24 \in A \cap B$ but $24 \notin C$.

3. Let $A = \{1, 3, ...99\}$ then $n(A) = 50$.
Let $B = \{1, 4, 9, ...100\}$ then $n(B) = 10$.
$n(A \cup B) = n(A) + n(B) - n(A \cap B) = 50 + 10 - 5 = 55$.

Yes, yes.

4. $6 * 6 = 36$. The cells of the partition are:
$A_1 = \{(1,1)\}, A_2 = \{(1,2),(2,1)\},$
$A_3 = \{(1,3),(2,2),(3,2)\} \cdots A_{11} = \{(6,6)\}.$

5. (a) $\{\{c\}, \{b\}, \{b,c\}, \{a\}, \{a,c\}, \{a,b\}, \{a,b,c\}, \{\ \}, \}$
(b) 2^k

6. (a) $\{(c,r) \mid 30 \le c \le 40 \text{ and } 10 \le r \le 15\}$
(b) $\{(c,r) \mid (20 \le c \le 29 \text{ and } 10 \le r \le 20) \text{ or } (30 \le c \le 40 \text{ and } 16 \le r \le 20)\}$

7. Is the barber in B? A universal set may be chosen to exclude the barber, and then the set is well-defined.

8. (a) function, (b) not a function.

9. (a) 10   (b) no, for example, not defined for $x = 2, y = 3$.
(c) the set of ordered pairs of natural numbers whose sum is even; i.e. x, y are both even or both odd.
(d) no. For example (3,5) and (6,2) both map to 4.

10. Yes, no. The function is not 1-1, for example (5,3) and (15,13) both map to 2. The function is 'onto' **N**.

11. (a) 2.6457513   (b) 14.09403   (c) 1096.6332   (d) 33.115452
Observe that $fog \ne gof$.

12. Each is 3.6. f and g are inverses of each other; $fog = gof =$ the identity function.

13. n is odd so $n + 1$ is even. So $(n + 1)/2$ is a whole number and the ceiling function does not change a whole number.

14. Any counterexample will do. For example $g(17) = g(9) = 1$.

15. Yes, yes, no, no.

16. (a) 3,5,2 (b) yes (c) yes (d) yes, $g^{-1} = (4x + 2) \bmod 7$ (e) yes

17. (a) 5,6,5 (b) yes (c) no (e.g. f(0) = f(4)) (d) no (e) no

18. (a) $51 - 6 = 5 \times 9$ (b) $124 - 7 = 13 \times 9$ (c) $175 - 13 = 18 \times 9$
(d) $u + t - (v + w) = (u - v) + (t - w) = km + lm = (k + l)m$
Therefore $u + t \equiv v + w \bmod m$
(e) $u = 100a + 10b + c = a + b + c + 99a + 9b \equiv abc \bmod 9$

19. (a) $27 - 5 = 2 \times 11$; $35 - 2 = 3 \times 11$ (b) $27 \times 35 - 5 \times 2 = 55 \times 11$

(c) $ut - vw = ut - vt + vt - vw = (u - v)t + v(t - w) = kmt + vlm =$ a multiple of m.

20. The relation must be reflexive, symmetric and transitive.

21. (a) The first point follows the second if it is on the same line and to the right or if it is on a line below.

(b) The relation is transitive.

If $\{[(c_1 > c_2) \wedge (r_1 = r_2)] \vee (r_1 > r_2)\}$ and $\{[(c_2 > c_3) \wedge (r_2 = r_3)] \vee (r_2 > r_3)\}$ then $\{[(c_1 > c_3) \wedge (r_1 = r_3)] \vee (r_1 > r_3)\}$.

22. Proof: $b = ma$ and $a = nb$ for some $m, n \in \mathbf{N}$.

Therefore $b = m(nb) = (mn)b$. Therefore $mn = 1$. Therefore $m = n = 1$ and $a = b$.

23. $a = b^n$ and $b = a^m$ for some $m, n \in \mathbf{N}$.

Therefore $a = (a^m)^n = a^{mn}$ and $mn = 1$. Therefore $m = n = 1$.

Therefore $a = b^1 = b$.

24. Transitive: If a divides b and b divides c then a divides c. Each element divides itself. For each pair of distinct elements of the set the 'left' divides the 'right'.

25. 2,4,6,8,10.

26. 0,1,3,7,15.

27. (a) $\{000, 001, 010, 100, 110, 101, 011, 111\}$ (b) `capable`

28. (a) 12, 17   (b) $1.41\dot{6}$ (c) 0.17346... 0.2% error

29. (a) $S_5 = \{5, 25, 35, 55, 65, 85, 95\}$, $S_{11} = \{11\}$ (b) 1

(c) $S_i \cap S_j = \emptyset$ since each number has a unique least prime factor; and every number has some least prime factor so belongs to one of the subsets A_k.

Exercise Set 4.4 (page 86)

1. (a)

p	q	$p \wedge q$	$\neg(p \wedge q)$	$p\|q$
T	T	T	F	F
T	F	F	T	T
F	T	F	T	T
F	F	F	T	T

(b) $p\|p$ is logically equivalent to $\neg(p \wedge p)$ is logically equivalent to $\neg p$.

(c)

$$\begin{aligned} p \wedge q &\equiv \neg\neg(p \wedge q) &&\text{(negation)} \\ &\equiv \neg(p\|q) &&\text{(defn. nand)} \\ &\equiv (p\|q)\|(p\|q) &&\text{(part (b))} \end{aligned}$$

(d)

$$\begin{aligned} p \vee q &\equiv \neg\neg(p \vee q) &&\text{(negation)} \\ &\equiv \neg(\neg p \wedge \neg q) &&\text{(de Morgan)} \\ &\equiv \neg p\|\neg q &&\text{(defn. of nand)} \\ &\equiv (p\|p)\|(q\|q) &&\text{(part (b))} \end{aligned}$$

2. Each has truth table T, F, T, T.

3.

x	y	$x \wedge y$	$\rightarrow$	x
T	T	T	T	T
T	F	F	T	
F	T	F	T	
F	F	F	T	

The proposition $x \wedge y \rightarrow x$ is true for all values of x, y. So the proposition is a tautology. So we write $x \wedge y \Rightarrow x$.

4. (a) A sufficient condition for 3 to divide n is that 9 should divide n.

A necessary condition for 9 to divide n is that 3 should divide n.

9 divides n only if 3 divides n.

(b) If $x = 0$ then $xy = 0$. $xy = 0$ is necessary for $x = 0$. $x = 0$ only if $xy = 0$.

5. (a) Converse: if n^2 is odd then n is odd. Contrapositive: if n^2 is even then n is even.

(b) If n is odd then n is prime and n is greater than 2. If n is even then n is not prime or $n \leq 2$.

(c) If $a = 0$ or $b = 0$ then $ab = 0$. If $a \neq 0$ and $b \neq 0$ then $ab \neq 0$.

6. Use truth table or

$$
\begin{aligned}
(p \lor q) \land (\neg p \lor q) &\equiv (p \land \neg p) \lor q &&\text{(distr., comm.)} \\
&\equiv \mathsf{F} \lor q &&\text{(contradiction)} \\
&\equiv q &&\text{(or-simplification).}
\end{aligned}
$$

7. $p \land (p \lor q) \equiv (p \lor \mathsf{F}) \land (p \lor q) \equiv p \lor (\mathsf{F} \land q) \equiv p \lor \mathsf{F} \equiv p$

8. (a)

$$
\begin{aligned}
(p \land q) \to (p \lor q) &\equiv \neg(p \land q) \lor (p \lor q) &&\text{(implication)} \\
&\equiv (\neg p \lor \neg q) \lor (p \lor q) &&\text{(de Morgan)} \\
&\equiv ((\neg p \lor \neg q) \lor p) \lor q &&\text{(associative)} \\
&\equiv (p \lor (\neg p \lor \neg q)) \lor q &&\text{(commutative)} \\
&\equiv ((p \lor \neg p) \lor \neg q) \lor q &&\text{(associative)} \\
&\equiv (\mathsf{T} \lor \neg q) \lor q &&\text{(excluded middle)} \\
&\equiv \mathsf{T} \lor q &&\text{(or-simpl., com.)} \\
&\equiv \mathsf{T} &&\text{(or-simpl., com.)}
\end{aligned}
$$

(b)

$$
\begin{aligned}
p \to (q \land r) &\equiv \neg p \lor (q \land r) &&\text{(implic.)} \\
&\equiv (\neg p \lor q) \land (\neg p \lor r) &&\text{(distr.)} \\
&\equiv (p \to q) \land (p \to r) &&\text{(implic.)}
\end{aligned}
$$

(c)

$$
\begin{aligned}
p \to (p \land q) &\equiv (p \to p) \land (p \to q) &&\text{(part (b))} \\
&\equiv \mathsf{T} \land (p \to q) &&\text{(implic.)} \\
&\equiv p \to q &&\text{(and-simplification)}
\end{aligned}
$$

(d) The proposition

$$
\begin{aligned}
&\equiv ((p \land q) \lor \neg p) \lor \neg q &&\text{(implication, de Morgan)} \\
&\equiv (p \land q) \lor \neg(p \land q) &&\text{(associative, de Morgan)} \\
&\equiv \mathsf{T} &&\text{(excluded middle)}
\end{aligned}
$$

(e)

$$
\begin{aligned}
((p \lor q) \land \neg p \to q &\equiv \neg((p \lor q) \land \neg p) \lor q &&\text{(implication)} \\
&\equiv (\neg(p \lor q) \lor p) \lor q &&\text{(de Morgan, neg.)} \\
&\equiv \neg(p \lor q) \lor (p \lor q) &&\text{(associative)} \\
&\equiv \mathsf{T} &&\text{(excluded middle)}
\end{aligned}
$$

9. (a)

$$
\begin{aligned}
\text{Statement} &\equiv \neg p \lor (\neg(q \lor r) \lor p) &&\text{(implication)} \\
&\equiv (\neg p \lor p) \lor (\neg(q \lor r)) &&\text{(assoc., comm.)} \\
&\equiv \mathsf{T} \lor (\ ) \equiv \mathsf{T}
\end{aligned}
$$

10. (a) T (b) T (c) T (d) T

11. True. The negation is $\exists_x \forall_y (y - x \neq x - y)$.

12.

$$\forall_{x \in \mathbf{R}, x > 0} \exists_{n \in \mathbf{N}} \, 2^n \leq x < 2^{n+1}$$

$$\exists_{x > 0} \forall_n \, x < 2^n \text{ or } x \geq 2^{n+1}$$

13. R, S. Truth set for P is $\{-2, -1, 1, 2\}$. Truth set for each of R, S is $\{1, 2\}$. Yes.

14. (a) T (b) T (c) p.

15. (a) $p \wedge (q \vee \neg r)$ (b) T.

Exercise Set 5.4 (page 105)

1. Proof: **Case 1.** $n = 3k$ (that is $n \equiv 0 \bmod 3$).
   The first factor is divisible by 3 so $n(n+1)(n+2)$ is divisible by 3.
   Case 2. $n = 3k + 1$ (that is $n \equiv 1 \bmod 3$).
   The third factor is $3k + 1 + 2 = 3(k+1)$ so $n(n+1)(n+2)$ is divisible by 3.
   Case 3. $n = 3k + 2$ (that is $n \equiv 2 \bmod 3$).
   The second factor is $3k + 2 + 1 = 3(k+1)$ so

   $n(n+1)(n+2)$ is divisible by 3. The three cases cover all possible integers so the proposition is true for all integers n.

2. Proof: The contrapositive of the proposition is,
   if m is even or n is even then mn is even.
   Case 1. m is even; then $m = 2k$ for some k.
   Thus $mn = (2k)n = 2(kn)$ which is even.
   Case 2. n is even; then $n = 2j$ for some j.
   Thus $mn = m(2j) = 2(mj)$ which is even. So mn is even if m is even or if n is even.

   Taking the contrapositive again we have the original proposition

   'if m is odd and n is odd then mn is odd'.

3. Proof: We are looking for a factor 3, so let us try to factorise $n^3 - n$.

$$
\begin{aligned}
n^3 - n &= n(n^2 - 1) \\
&= n(n-1)(n+1) \\
&= (n-1)n(n+1)
\end{aligned}
$$

which is the product of three consecutive numbers. One of these three numbers must be divisible by 3 (proven in question 1), so the whole number is divisible by 3. That is $n^3 - n$ is divisible by 3.

4. Proof: Suppose the given proposition is false. Then we must have

$$ab = n \ (1) \text{ and } a < \sqrt{n} \text{ and } b \le \sqrt{n}$$

Now, because a, b are positive and $a < \sqrt{n}$ and $b \le \sqrt{n}$ it follows that

$$(a)(b) < (\sqrt{n})(\sqrt{n}),$$

that is

$$ab < n$$

which contradicts (1) above.

So our supposition in the first line of this proof was false, which means the original proposition is true.

5. Proof: Define the predicate $P(n) = (n^2 > 2n + 1)$.
   Basic: Show that $P(3)$ is true.

$$3^2 = 9.2 * 3 + 1 = 7 \text{ so } 3^2 > 2 * 3 + 1$$

Therefore $P(3)$ has value T.
Inductive: Show that $(P(k) \to P(k+1))$ is true for all integers $k \ge 3$.

Assume that $P(k)$ is true. Thus $k^2 > 2k + 1$ for a particular constant value k.

$$
\begin{aligned}
\text{Thus } (k + 1)^2 \ &= \ k^2 + 2k + 1 \\
&> \ (2k + 1) + 2k + 1 \\
&> \ 2k + 1 + 1 + 1 \text{ since } k \ge 3 \\
\text{thus } (k + 1)^2 \ &> \ 2(k + 1) + 1 \\
&\quad \text{which is } P(k + 1)
\end{aligned}
$$

We have shown that $(P(k) \to P(k+1))$ is true for a particular constant k. But the argument is quite general and we deduce that $(P(k) \to P(k+1))$ for every $k \ge 3$ So by the principle of mathematical induction $n^2 > 2n + 1$ for all integers $n \ge 3$.

6. After line 1 we have $(a = X, b = Y, temp = Y)$.
   After line 2, $(a = X, b = X, temp = Y)$.
   After line 3, $(a = Y, b = X, temp = Y)$

and the last set of values satisfies Q, so we have proved the algorithm correct.

7. (a) $(5 \geq 1 + 4)$. Yes.

   (b) $5^k \geq 1 + 4k$

   (c) $5^{k+1} \geq 1 + 4(k+1)$

   (d) $5^{k+1} = 5 \times 5^k \geq 5 \times (1 + 4k)$, given.

   Thus $5^{k+1} \geq 1 + 4k + 4 + 16k > 1 + 4(k+1)$ since $k > 0$.

   Therefore $P(k+1)$ is true.

8. (a) $a \in \{1, 2, \ldots 9\}, \quad b, c \in \{0, 1, \ldots 9\}$.

   (b) $n = 100a + 10b + c$.

   (c) Proof: $n = 100a + 10b + c = 100a + 10(a + c) + c = 110a + 1c$

   So $n = 11(10a + c)$ which is divisible by 11.

   (d) No. For example 517 is divisible by 11 but $5 + 7 \neq 1$.

   (e) Different generalisations are possible, to cover all cases of divisibility by 11 and numbers of any length. I liked the answer of the student who generalised to binary representation.

9. Mathematical induction. Define the predicate $P(n)$ to be $(n! > 2^n)$.

10. Mathematical induction. Define the predicate $P(n)$ to be $(2^n > n^2)$.

11. Mathematical induction. Define the predicate $P(n)$ to be

$$a + (a + d) + (a + 2d) + \cdots + (a + (n-1)d) = \frac{n}{2}[2a + (n-1)d]$$

Basic: $P(1) \equiv (a = \frac{1}{2}[2a + (1-1)d])$ is true.

Inductive: Assuming the formula for k terms is correct we have:

The sum of the first $k+1$ terms

$$= \frac{k}{2}[2a + (k-1)d] + a + kd$$

$$= \frac{2ak + k(k-1)d}{2} + \frac{2a + 2kd}{2}$$

$$= \frac{2a(k+1) + kd(k-1+2)}{2}$$

$$= \frac{(k+1)}{2}[2a + ((k+1) - 1)d]$$

which is $P(k+1)$.

12. Mathematical induction. Define $P(n)$ to be $(S_n = 2^n - 1)$.

Basic: $S_0 = 0$ (given). $2^0 - 1 = 1 - 1 = 0$.
Therefore $S_n = 2^n - 1$ for $n = 0$, and $P(0)$ is true.

Inductive:

$$
\begin{aligned}
S_{k+1} &= 2S_k + 1 && \text{(definition of } S_n\text{)} \\
&= 2(2^k - 1) + 1 && \text{(assuming P}(k)\text{)} \\
&= 2^{k+1} - 1 && \text{(algebra)} \\
S_{k+1} &= 2^{k+1} - 1
\end{aligned}
$$

which is $P(k + 1)$. So by the principle of mathematical induction $S_n = 2^n - 1$ for all $n \geq 0$.

13. Proof by cases.

14. Assume $P(k)$ (i.e. assume $n^2 + n + 1 = 2m$ for some m).

$$
\begin{aligned}
(k + 1)^2 + (k + 1) + 1 &= k^2 + 2k + 1 + k + 1 + 1 \\
&= (k^2 + 2k + 1) + 2k + 2 \\
&= 2m + 2k + 2
\end{aligned}
$$

which is even. Thus $P(k) \to P(k+1)$, and k is an *arbitrary* constant, so in fact the result holds for all integers k.

However $P(n)$ is false for all n. We may write

$$n^2 + n + 1 = n(n + 1) + 1$$

that is the product of consecutive numbers plus 1. But one of the consecutive numbers must be even, so the product is even, and on adding 1 the sum must be odd. The significance of this question is that for a proof by mathematical induction *both* the basic and the inductive parts of the argument are necessary.

15. Take out the common factor $(-1)^n$ on the left-hand side of the equation.

Exercise Set 6.4 (page 127)

1. (a) 10   (b) 7   (c) 0

2. *product* $: \mathbf{N} \to$ *Codomain of f*

$$
product(n) = \begin{cases} 1 & \text{if } n \leq a \\ f(n) \times product(n - 1) & \text{otherwise} \end{cases}
$$

3. (a) 14400

(b) $n^2 \times (n-1)^2 \times \cdots \times 2^2 \times 1^2 = (n \times (n-1) \times \cdots \times 2 \times 1)^2 = (n!)^2.$

(c) 1

4. $a_1 = 2, a_2 = 3.$ For $n > 2,$

$a_n = a_{n-1}(\text{ending 'white'}) + a_{n-2}(\text{ending 'white red'}).$

5. (a) For $n \geq 2, s_n = s_{n-1} + 1; s_1 = 2.$

(b) $s_n = n + 1.$

6. $s_1 = 3, s_2 = 8.$

For $n \geq 2, s_n = 2s_{n-1}(\text{ending in 0 or 2}) + 2s_{n-2}(\text{ending 01 or 21}).$

7. The recursive formula is:

$$d_n = \begin{cases} d_{n-1} * (1.0125) - 500, & \text{if } n \geq 1 \\ 30000 & \text{if } n = 0. \end{cases}$$

8. (a) $s_0 = 5000; s_j = (1.01) * s_{j-1} + 200$ for $1 \leq j \leq 18$

(b)

$$\begin{aligned} s_{18} &= 5000 * (1.01)^{18} \\ &+ 200(1 + 1.01 + (1.01)^2 + \cdots + (1.01)^{17}) \\ &= 5000 * (1.01)^{18} + \frac{200((1.01^{18}) - 1)}{1.01 - 1} \\ &= 9903.69 \end{aligned}$$

9. $x_{50}, x_{25}, x_{12}, x_{18}$

10. The seventh item in the list is 12.49 ($\sqrt{156}$)

11. $< v, n >$. The list $remove(x,l)$ is produced by removing each occurrence of the item x from the list l.

$$\begin{aligned} \text{for example} \quad & remove(e, < e, v, e, n >) \\ & remove(e, < v, e, n >) \\ =\ & < v > \| \, remove(e, < e, n >) \\ & remove(e, < n >) \\ =\ & < n > \| \, remove(e, <>) \\ & <> \\ =\ & < v > \| < n > \| <> \\ =\ & < v, n > \end{aligned}$$

12. $removefirst : Char \times List \to List$

$removefirst(x, l)$

$$= \begin{cases} l & \text{if } l = <> \\ tail(l) & \text{if } head(l) = x \\ < head(l) > || removefirst(x, tail(l)) & \text{otherwise} \end{cases}$$

13. $replace : Char \times Char \times List \to List$

$replace(x, y, l)$

$$= \begin{cases} l, & \text{if } l = <> \\ < y > || replace(x, y, tail(l)), & \text{if } head(l) = x \\ < head(l) > || replace(x, y, tail(l)), & \text{otherwise} \end{cases}$$

Note: Every occurrence of x is replaced by y.

14.

$$isEven : \quad \mathbf{N} \to B$$
$$isEven(n) \quad = \quad \begin{cases} \mathsf{T} & \text{if } n = 0 \\ \mathsf{F} & \text{if } n = 1 \\ isEven(n-2) & \text{otherwise} \end{cases}$$

15. This could be done using 'lists' as a model. First define a function *tail* on strings in A^+, that removes the first character from a non-empty list. Then:

$$length : \quad A^* \to \mathbf{N}$$
$$length(\alpha) \quad = \quad \begin{cases} 0 & \text{if } \alpha \text{ is empty} \\ 1 + length(tail(\alpha)) & \text{otherwise} \end{cases}$$

16.

$$power: \quad \mathbf{N} \times \mathbf{N} \to \mathbf{N}$$
$$power\,(a, n) = \begin{cases} 1 & \text{if } n = 0 \\ a * power(a, n-1) & \text{if } n > 0 \end{cases}$$

17. $x_1 = 1, x_2 = 2$. For $n > 2, x_n = x_{n-1} + x_{n-2}$.

18. (a) method: $xone \leftarrow 1$
$xtwo \leftarrow 3$
for $j = 3$ to 200
$$\begin{cases} x & \leftarrow & 3 * xtwo - 2 * xone \\ xone & \leftarrow & xtwo \\ xtwo & \leftarrow & x \end{cases}$$

(b) $x_n = 2^n - 1$

19. $s_n = 5 * 3^n, n \geq 0$

20. $x_n = 2 * 3^n - 1$

21. $s_n = 3 + (-2)^n, n \geq 0$

22. $a_n = 2^n + 3 * 4^n$

23. $a_n = 2 * 3^n$

24. $5^n(n-2(n-1)+(n-2)) = 0$. The general solution is $s_n = c_1 5^n + c_2 n 5^n$.

25. $a_n = (-3)^n(3 - 5n)$

26. (a) Conjecture $a_n = 2n + 1, \quad n \geq 0$ (b) Conjecture is correct

27. (a) $a_n = n + 1$

28. $s_n = s_{n-1} + s_{n-2} \quad n \geq 3. s_1 = 1, s_2 = 2$.

$$s_n = \frac{3 + \sqrt{5}}{2\sqrt{5}}\left(\frac{1 + \sqrt{5}}{2}\right)^{n-1} - \frac{3 - \sqrt{5}}{2\sqrt{5}}\left(\frac{1 - \sqrt{5}}{2}\right)^{n-1}$$

29. $s_1 = 3, s_2 = 7; s_n = 2(s_{n-1} - s_{n-2}) + 3s_{n-2} = 2s_{n-1} + s_{n-2}, n \geq 3$

30. $s_n = \frac{n(n-1)}{2}$. The order is n^2.

31.
$$s_n = \frac{1 + i}{2}(-1 + i)^n + \frac{1 - i}{2}(-1 - i)^n, \quad n \geq 0.$$

Note: s_n itself is not complex. The formula generates the sequence $1, -2, 2, 0, -4, \cdots$.

Exercise Set 7.4 (page 150)

1. (a) 1524 (b) 1 050 576 (c) $1.0995 * 10^{12}$ (d) $1.2089 * 10^{24}$

2. (a) 85.4166... Algorithm 2 is about 85 times faster than algorithm 1 for a problem in which $n = 1024$.
(b) 21.8 seconds.

3. (a) $T_1(5000) = 2607, T_2(5000) = 9000$. Prefer algorithm 1.
(b) 2.896. Say 3 seconds (c) 17.5 seconds

4. 2.285

5. (a) 45057   (b) 3.9 seconds   (c) Algorithm 2 is about 57 times slower than algorithm 1 for $n = 1024$.

6. $n < n \log_2 n < n^{\frac{3}{2}} < n^2$

7. (a) n^2 (b) n^2 (c) $n \log_2 n$ (d) 2^n (e) $n!$

8. The algorithms are each of order n since, taking 'comparison of two numbers' as the unit of work, each has time complexity $n - 1$.

9. 120 minutes, $0 - 2 - 1 - 5 - 3 - 4 - 0$.

10. The minimum cost is \$2800. The sequence $4 - 3 - 1 - 2 - 5$ is one of two sequences that costs least.

11.   (a) $\gcd(21, 34) = \gcd(13, 21) = \cdots = \gcd(0, 1) = 1$ (8 function calls); $\gcd(20, 205) = 5$ (3 function calls); $\gcd(23, 37) = 1$ (7 function calls). The number of function calls is large when the gcd is 1. The worst case is for the input 21, 34, a pair of Fibonacci numbers.

   (b) For $n \geq 2$ we have $F_{n+1} = F_n + F_{n-1}$ and from chapter 6 (p. 126) it follows that $0 < F_{n-1} < F_n$. Thus F_{n-1} is the remainder on dividing F_{n+1} by F_n. Therefore $\gcd(F_n, F_{n+1}) = \gcd(F_{n-1}, F_n)$.

   (c) $\gcd(F_n, F_{n+1}) = \gcd(F_{n-1}, F_n) \cdots \gcd(0, F_1) = 1$ requires $n + 1$ function calls (counting the subscripts of the second member of each pair).

   (d) For the time complexity of the algorithm, a may be taken to represent the 'size' of the problem. Execution time may be taken to be proportional to the number of times the function gcd is called.

   The 'worst' case, among problems of roughly equal size, is that for which the most function calls are required; and that happens when the gcd is 1 and successive remainders in the calculation decrease by as little as possible.

   Therefore, the last function call must have $a = 0$ and $b = 1$, yielding $\gcd(0, 1) = 1$. For $a \neq 0$, successive function calls will be

   1. $\gcd(b, \ )$
   2. $\gcd(a, b)$
   3. $\gcd(x, a)$

where $b = aq + x$ and $0 \le x < a$ for some integer $q > 0$. In general, b, x are remainders and we would like $b - x$ to be as small as possible.

For $x \ne 0$, $b - x = aq$ is a minimum for $q = 1$. In this case

$$b = x + a$$

For $x = 0$, we must have $a = 1$ and $b = q$ is a minimum for $q = 2$ (since $b > a$). Thus

$$1 = \gcd(0,1) = \gcd(1,2) = \gcd(2,3) = \gcd(3,5) = \cdots$$

where $1, 2, 3, 5, \cdots$ are the Fibonacci numbers $F_1, F_2, F_3, \cdots$ generated by $F_1 = 1$ (i.e. $a = 1$), $F_2 = 2$ (i.e. $b = 2$) and, for $n > 2$, $F_n = F_{n-1} + F_{n-2}$ (i.e. $b = x + a, x \ne 0$).

Thus the worst case for the evaluation of $\gcd(a, b)$ occurs when a, b are successive Fibonacci numbers.

(e) The worst case time complexity function for $\gcd(a, b)$ is (from above) $T(a) = n + 1$ where $a = F_n$, $b = F_{n+1}$.

Given $a \approx \dfrac{1}{\sqrt{5}} \left(\dfrac{1 = \sqrt{5}}{2} \right)^{n+1}$ we find

$$T(a) = n + 1 = \frac{\log(\sqrt{5}a)}{\log \frac{1+\sqrt{5}}{2}}$$

(f) The number of calls $T(832\,040) = 30$. The gcd is 1.

12. $< \cos 50, \cos 100, \cos 200, \cos 300 > = < .64, -.17, -.94, 0.5 >$
$= merge(mergesort(< \cos 50, \cos 100, >,$
$\qquad\qquad\qquad mergesort(< \cos 200, \cos 300 >))$
$= merge(merge(< \cos 50 >, < \cos 100, >),$
$\qquad\qquad\qquad merge(< \cos 200 >, < \cos 300 >))$
$= merge(< \cos 100, \cos 50, >, < \cos 200, \cos 300 >)$
$= < \cos 200, \cos 100, \cos 300, \cos 50 >$

13. $< \cos 3, \cos 4, \cos 2, \cos 5, \cos 1 >$

14. For investigation.

Exercise Set 8.4 (page 183)

1. (a) Yes. Exactly two vertices of odd degree.

(b) Level(0): v_1, Level(1): v_3, v_5, Level(2): v_6, v_4, Level(3): v_2.

(c) v_2 (d) $v_1 v_3 v_5 v_4 v_2 v_6$

2. (a) No   (b) Yes e.g. 412653 or 352614   (c) $\begin{array}{ccccc} 4 & 1 & 2 & 5 & 3 \\ & & 6 & & \end{array}$; v_3

(d) $\begin{array}{ccccc} 4 & 1 & 2 & 5 & 3 \\ & & 6 & & \end{array}$

3. (a) vertex subscripts 1351634624

(b) The tree contains the edges (4,2), (4,3), (4,6), (3,1), (3,5).

(c) v_1, v_5

4. Let the number of sailors on the left bank be x.
Let the number of pirates on the left bank be y.
Let the location of the boat be z.
Then $0 \le x \le 3$; $0 \le y \le 3$; $z \in \{L, R\}$ and $x \ge y \lor x = 0$.

Represent the states (x, y, z) by the vertices of a graph and, if it is possible to transfer between two states by a single boat crossing join the corresponding vertices by an edge.

For example $(2, 2, L)(1, 1, R)$ is an edge; $(3, 1, L)(2, 2, R)$ is not an edge.

A path which solves the problem is:
$(3, 3, L)(3, 1, R)(3, 2, L)(3, 0, R)(3, 1, L)(1, 1, R)$
$(2, 2, L)(0, 2, R)(0, 3, L)(0, 1, R)(0, 2, L)(0, 0, R)$.

Other paths are possible.

5.

$$A^2 = \begin{pmatrix} 2 & 0 & 0 & 2 \\ 0 & 2 & 2 & 0 \\ 0 & 2 & 2 & 0 \\ 2 & 0 & 0 & 2 \end{pmatrix}$$

6. \$39 000

Exercise Set 9.5 (page 216)

1. 128.

2. $n - (\lfloor \frac{n}{2} \rfloor - 1) = n/2 + 1$ if n is even; $n/2 + 3/2$ if n is odd.

3. 153

4. (a) 900 (b) 450 (c) 648 (d) 320

5. 36

6. (a) 4 (b) 2^{k-1}

7. 833

8. 250

9. (a) 500 (b) 96 (c) 60 (d) 48 (e) 42

10. 30

11. $\frac{n(n-1)}{2}$

12. mn

13. 488

14. 600

15. For discussion.

16. There are 24 sequences of 4 numbers with total sum $4 \times 300 = 1200$. The average sum is 50 . But the sequences cannot all be equal so at least one sequence has a sum greater than 50.

17. (a) 30 (b) 1 663 200 (c) 16 216 200

18. (a) 1000 (b) 280 (c) 0.28

19. 120

20. 210

21. 11 975 040

22. 136

23. 495

24. 56

25. 96

26. Every number from 1 to 300 may be expressed uniquely as an odd number (including 1) by a power of 2 (including 2^0). Thus $S = \{1, 2, 3, \cdots, 300\}$ may be partitioned into 150 subsets

$$S_a = \{2a^i \,|\, a \text{ is odd}, i \in \mathbf{N}\}$$

one subset for each odd number $a = 1, 3, 5, \cdots 299$. If we choose 151 numbers from S, two must be chosen from the same subset S_a. But the smaller of these must divide the larger (e.g. choosing $a2^i, a2^j$ from S_a, if $i < j$ then $a2^j \div a2^i = 2^{j-i}$), and the result follows.

Exercise Set 10.4 (page 246)

1. Counterexample:$64 \div (8 \div 2) = 16, (64 \div 8) \div 2 = 4.4 \neq 16$.

2. concatenate(concatenate(com,put),er)
   = concatenate(comput,er) = computer
   concatenate(com, concatenate(put, er)
   = concatenate(com,puter) = computer

3. For all $X \in \mathcal{P}(A)$, $\quad X \cap X = X$.

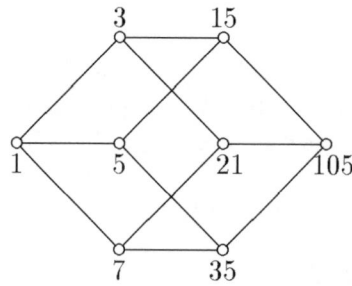

Figure B.1: Hasse diagram Ex. 4

4. 21, 105, 21. The least element is 1, greatest element 105.

   Define complement by $complement(a) = a' = \frac{105}{a}, a \in L$.

5. (a) i. $\forall x, y \in \mathbf{B}x \wedge y = y \wedge x$,
       ii. $\forall x, y, z \in \mathbf{B}x \vee (y \vee z) = (x \vee y) \vee z$
       iii. $\forall x, y, z \in \mathbf{B}x \vee (y \wedge z) = (x \vee y) \wedge (x \vee z)$

   (b) T is the identity for $\wedge$.

   (c) F is the identity for $\vee$.

   (d) There is no inverse for T.

6. (a) $\begin{pmatrix} 0 & 0 \\ 0 & 0 \end{pmatrix}$ (b) yes (c) yes (d) yes (e) no (f) yes to each

7. 'append' is associative.

8. (a) yes   (b) no   (c) no, $c * c = b$ (d) yes, b

(e) c doesn't have a unique inverse.

12. Case 2. $m < 0$

$$
\begin{aligned}
(p^m)^{-1} &= ((p^{-m})^{-1})^{-1} & \text{(definition 10.14)} \\
&= p^{-m} & \text{(theorem 10.3)} \\
\text{and} \quad (p^{-1})^m &= ((p^{-1})^{-m})^{-1} & \text{(definition 10.14)} \\
&= ((p^{-m})^{-1})^{-1} & \text{(case 1. } m > 0.) \\
&= p^{-m} & \text{(theorem 10.3)} \\
\text{Therefore} \quad (p^m)^{-1} &= p^{-m}
\end{aligned}
$$

16. (a)

$$
\begin{aligned}
x.0 &= x.(x.\overline{x}) & \text{(definition 10.21)} \\
&= (x.x).\overline{x} & \text{(. is associative)} \\
&= x.\overline{x} & \text{(. is idempotent)} \\
&= 0 & \text{(definition 10.21)}
\end{aligned}
$$

17. (a)

$$
\begin{aligned}
x.(x + y) &= (x + 0).(x + y) & \text{(0 is the identity for +)} \\
&= x + (0.y) & \text{(+ distributive over .)} \\
&= x + 0 & \text{(exercise 16(a))} \\
&= x & \text{(0 is the identity for +)}
\end{aligned}
$$

18. (a) Proof:

$$
\begin{aligned}
(x.y) + (\overline{x} + \overline{y}) &= (x.y + \overline{x}) + \overline{y} & \text{(+ is associative)} \\
&= (\overline{x}x.y) + \overline{y} & \text{(+ commutative)} \\
&= (\overline{x} + x).(\overline{x} + y) + \overline{y} & \text{(+ distrib. over .)} \\
&= (\overline{x}x.y) + \overline{y} & \text{(+ commutative)} \\
&= \overline{x} + 1 & \text{(def 10.21)} \\
&= 1 & \text{(exercise 16(b))} \\
\overline{x.y} &= \overline{x} + \overline{y} & \text{(def 10.21)}
\end{aligned}
$$

Exercise Set 11.3 (page 268)

1. (1) premise 1 (2) premise 2 (3) 1, ∧-E (4) 3,2, →-E (5) 1, ∧-E
(6) 4,5, ∧-I

2. (1) premise 1 (2) 1, ∧-E (3) premise 2 (4) 2,3, →-E (5) 4, ∨-I

3. (a) ∧-E   (b) premise 1, premise 3, →-E

(c) premise 1, ∨-I   (d) ¬-E

4. *Note:* The following is part of a proof of one of de Morgan's laws.

   1. $\neg(p \lor q)$ assumption for →-I
    2. p assumption for ¬-I
    3. $p \lor q$ 2, ∨-I
    4. $\neg(p \lor q) \land (p \lor q)$ 1,3, ∧-I
   5. $\neg p$ 2,4, ¬-I
    6. q assumption for ¬-I
    7. $p \lor q$ 6, ∨-I
    8. $\neg(p \lor q) \land (p \lor q)$ 1,7, ∧-I
   9. $\neg q$ 6,8, ¬-I
  10. $\neg p \land \neg q$ 5,9, ∧-I
  11. $\neg(p \lor q \to \neg p \land \neg q)$ 1,10, →-I

5. (a) 1. $p \land (p \to q)$ assumption for →-I
      2. p 1, ∧-E
      3. $p \to q$ 1, ∧-E
      4. q 2,3, →-E
      5. $p \land (p \to q) \to q$ 1,4 →-I

  (b)1. $\neg q$ premise 1
      2. q assumption for →-I
      3. $q \lor p$ 2, ∨-I
      4. p 1,3, example 2
     5. $q \to p$ 2,4 →-I

A proof could have been based on the law of implication (example 5).

6.

$$
\begin{aligned}
x + 1 &= x + (x + \overline{x}) && \text{(complement property)} \\
&= (x + x) + \overline{x} && \text{(associative property)} \\
&= x + \overline{x} && \text{(idempotent property)} \\
&= 1 && \text{(complement property)}
\end{aligned}
$$

The proof uses the fact that $x, x, \overline{x}, 1$ are elements of the given Boolean algebra. The elements of a Boolean algebra satisfy the properties stated at the end of each line. In each case the rule of inference used is →-I, modus ponens.

7. (a) $a \leq 1$ (b) $a + b = 0$ (c) $(a = -2) \lor (b = 2)$
(d) $zj^i = c$

(e) $wp(A,Q) \equiv (n$ is even $\wedge\; wp(n \leftarrow n/2; y \leftarrow y^2, (y^n = c))) \vee (n$ is odd $\wedge (y^n = c)) \equiv (y^n = c)$

10.

$$
\begin{aligned}
wp(A,Q) &\equiv wp(i \leftarrow i+1; s \leftarrow s+i, (s = \frac{i(i+1)}{2})) \\
&\equiv wp(i \leftarrow i+1, s+i = \frac{i(i+1)}{2}) \\
&\equiv (s+i+1 = \frac{(i+1)(i+2)}{2}) \\
&\equiv (s = \frac{(i+1)(i+2)}{2} - \frac{2(i+1)}{2}) \\
&\equiv (s = \frac{i(i+1)}{2}) \quad \equiv Q \quad \equiv P
\end{aligned}
$$

12. Let k be the number of executions of the body of the loop remaining before termination.

If $k = 0$ then $i \geq n$ and Q must hold, i.e. $fac = n!$

$$P_0 \equiv ((i \geq n) \wedge (fac = n!))$$

If $k = 1$ then $i < n$ and the postcondition for an execution of the body of the loop is P_0.

$$
\begin{aligned}
P_1 &\equiv (i < n) \wedge wp(i \leftarrow i+1, wp(fac \leftarrow fac \times i, P_0)) \\
&\equiv (i < n) \wedge wp(i \leftarrow i+1, (i \geq n) \wedge (fac \times i = n!)) \\
&\equiv (i < n) \wedge (i+1 \geq n) \wedge (fac \times i+1 = n!)) \\
&\equiv (i = n-1) \wedge (fac = (n-1)!)
\end{aligned}
$$

In general
$$P_k \equiv (i = n-k) \wedge (fac = (n-k)! = i!)$$

Now P_k is the precondition that guarantees Q after exactly k executions of the body of the loop. Therefore

$$wp(\text{while loop}, (fac = n!)) \equiv \exists_{k \geq 0}, (i = n-k) \wedge (fac = i!)$$

Choosing $k = n$ we obtain the weakest precondition $P \equiv (i = 0) \wedge (fac = 0! = 1)$. An invariant for the loop is $fac = i!$.

Bibliography

[1] Aho, A.V., Hopcroft, J.E. and Ullman, J.D., (1984), *The Design and Analysis of Computer Algorithms*, Addison-Wesley.

[2] Backhouse, R.G., (1986), *Program Construction and Verification*, Prentice-Hall.

[3] Birkhoff, G. and MacLane, S. (1953), *A Survey of Modern Algebra*, rev. edn, MacMillan.

[4] Cooke, D.J. and Bez, H.E., (1984), *Computer Mathematics*, Cambridge University Press.

[5] Davis, P.J. and Hersh, R., (1981), *The Mathematical Experience*, Harvester.

[6] Goodstein, R.L., (1957), *Recursive Number Theory*, North Holland.

[7] Gries, D., (1981), *The Science of Programming*, Springer-Verlag.

[8] Grogono, P. and Nelson, S., (1982), *Problem Solving and Computer Programming*, Addison-Wesley.

[9] Halmos, P.R., (1961), *Naive Set Theory*, Van Nostrand.

[10] Knuth, D.E., (1969), *The Art of Computer Programming, Vol. 1, Fundamental Algorithms*, rev. edn, Addison-Wesley.

[11] Slater, G.(ed.), (1987), *Essential Mathematics for Software Engineers*, The Institution of Electrical Engineers (UK).

Index

$\emptyset$, the empty set, 42
$\exists$, there exists, 83
$\forall$, for all, 83
$\leftarrow$, assign a new value, 22
$\leftrightarrow$, if and only if, 77
$\neg$, not, 71
$\rightarrow$, if $-$ then, 75
$\subseteq$ is a subset of, 43
$\underline{\vee}$, exclusive or, 75
$\vee$, or, 66, 74
$\wedge$, and, 66, 72
$\{P\}A\{Q\}$, 261

Abel, Niels Henrik, 230
Abelian , 234
Abelian group, 230
absolute error, 35
absolute value, 51
absorption, 242
addition
    of integers, 22
    of rationals, 30
adjacency
    list, 162
    matrix, 162
adjacent, 156
algebra, 222, 229
algorithm, 1, 50
    correctness of, 8
    Euclid's, 11
    prime factors, 9
    prime test, 8
    termination of, 12
alphabet, 42, 61

and (conjunction), 72
antisymmetric, 58
append, 119
arbitrary, 254
Archimedes, 38
arithmetic sequence, 120
arithmetic series, 107
arrangement, 210
assignment, 272
associative, 224
atomic, 70
axiom, 78

Babylonian, 38
base, 17–25
binary, 17
binary directed ordered tree (BDOT),
        172–175
binary information tree, 175
binary operation, 222
binary relation, 54
binary search, 116, 144
binary string, 18
binomial theorem, 203–205
bipartite graph, 185
bit, 26
body of loop, 266, 273
Boole, George, 69, 249
Boolean, 4, 36
Boolean algebra, 240
branch, 265, 273
bridge, 167
bubblesort, 145–147, 153

calculus, 78
Cartesian product, 46
cell of a partition, 47
changing base, 19
character, 42, 61
characteristic function, 43, 207
circuit, 164
clashes, 254
closed, 221
codomain, 48
combinatorial reasoning, 195, 199–
    206
commutative, 224
complement, 192, 223, 242
complement of a set, 45
complemented lattice, 242
complete bipartite graph, 217
complete graph (K_n), 217
complexity function, 136
component of a graph, 169
composite integer, 7
composition of functions, 52
compound proposition, 70
concatenate, 62
concatenation, 223
congruence modulo m, 14, 55
conjunctive normal form, 88
connected graph, 167
connective, 70
    logical, 70
    meaning of, 71
consecutive numbers, 194
contingency, 79
contradiction, 73, 79, 258
contrapositive, 76
converse, 76
correctness, 146, 260
counting
    r-combinations, 196
    r-multisets, 197
    r-permutations, 196

r-sequences, 195
r-subsets, 196
    empty subsets, 198
    sequences with repetition, 200
counting walks, 166
cycle in a graph, 164
cyclic group, 234
cycling, 32, 34

Davis, Philip, 15
De Morgan, A., 74, 242
decision problem, 7
degree of a vertex, 159
depth in a search tree, 177
dialectic, 1
digraph, 158
discharged, 252
disconnected, 169
displacement sequence, 211
distributive, 228
distributive lattice, 242
divisible, 7
division
    of integers, 24
    of rationals, 31
division theorem, 3
domain, 49
done, 177

edge (arc), 155
edge label, 159
efficiency of an algorithm, 122
element of a list, 118
element of a set, 41
empty list, 118
empty vertex, 176
equality
    of rational numbers, 31
    of sets, 44
equivalence class, 57
equivalence relation, 57, 169
equivalence relations, 56

errors, 34
Euclid, 10, 89
Euclid's algorithm, 141
Euler walk, 167
Euler, Leonhard, 167
exclusive or, 74, 75
execution properties, 133
exponential time, 140
exponentiation, 233

factorial, 110, 198
Fibonacci, 127
Fibonacci numbers, 126, 153
field, 244
fixed length word, 26
Fleury's algorithm, 168
floor, 8
for loop, 274
formal proof, 257
free, 255
function
    properties of, 50
fundamental theorem of arithmetic,
    7
fuzzy set, 48

gcd algorithm
    worst case, 152
generalisation, 255, 258
generalised associative, 229
generalised distributive, 229
generating
    permutations, 207, 212
    subsets, 207
Gentzen, Gerhard, 250, 268
geometric sequence, 120
graph
    general graph, 156
    multigraph, 156
    null, 156
    simple graph, 156
Gray code, 208

greatest common divisor, 11
greatest element, 240
greatest lower bound (glb), 238
group, 230, 256
    Abelian or commutative, 232
    properties, 232

Halmos, Paul, 58
Hamiltonian
    cycle, 169, 209
    path, 169
Hasse diagram, 237
Hasse, Helmut, 237
head of a list, 119
Hersh, Reuben, 15
hexadecimal, 17
Horner's algorithm, 21, 34, 137

idempotent, 225
identity, 226
identity function, 53
if − then (conditional), 75
if and only if (biconditional), 77
image, 50
incidence function, 156
incident, 155
inclusion/exclusion, 192
inclusive or, 73
induced partial order, 243
inference, 250
infix notation, 222
insert, 120
instantiation, 255
instructions, 271
integers, 2
    closed under addition, 25
    closed under multiplication, 25
    division of, 3–7
intersection of sets, 44
intractable, 140
invariant, 12, 262, 267
inverse, 76, 227

inverse function, 53
irrational numbers, 30, 35
isomorphism, 235
iterative, 122

job sequencing, 136, 152
juxtaposed, 61

Königsberg bridge problem, 167
Kleene closure, 62
Knuth, Donald, 15
Kruskal's algorithm, 181

last in first out, 179
lattice, 239
law of contradiction, 80
law of excluded middle, 252
law of identity, 252
law of implication, 80, 253
law of the exclude middle, 80
laws of logic, 79–81
least element, 240
least upper bound (lub), 239
Leibniz, 69
length of a list, 119
length of a string, 61
level, in a search tree, 177
linear search, 143
list, 60, 118
literal expansion, 18, 111
logically equivalent, 72, 79
loop, 4, 156, 273
loop invariant, 13

mapping, 50
mathematical induction, 94, 259
    strong form, 101
matrix, 160
matrix multiplication, 161
meaning of logical connectives, 71,
        73
merge, 147

mergesort, 147–150, 153
minimal spanning tree, 181
model, 157
modus ponens, 100, 251, 259
Mohammed al-Khwarismi, 2
monoid, 230
Morse code, 217
multinomial theorem, 205–207
multiple edges, 156
multiplication
    of integers, 23
    of rationals, 31
multiset, 42

nand, 86
Nassi-Schneidermann, 9
natural deduction, 250
necessary, 77
negation, 85
negative exponentiation, 233
negative integers, 26
Newton, Isaac, 37
not (negation), 71
null tree, 174

octal, 17
one-one, 51
one-one function, 189
onto, 51
operation
    binary, 54
    unary, 54
operation count, 137
operation symbol (meaning), 271
or (disjunction), 74
order of a function, $O(g(n))$, 139
order of bubblesort, 146
order of quantifiers, 85
ordered field, 245
ordered set, 59
overflow, 28

paradox, 187
   barber, 63
   Russell's, 47
parent, 176
partial function, 48, 49
partial order, 58
partially ordered set, 236
partition of a set, 47, 57, 170, 192
partitioned matrix, 185
path, 163
permutation, 211
pigeonhole principle, 193
pointer, 177
polynomial time, 140
positive integer division, 3
postcondition, 5, 83, 261
postfix notation, 222
power set, 43
precedence, 77
precondition, 5, 83, 261, 262
predecessor, 52, 61, 176
predicate, 42, 83
prefix notation, 222
prime, 7
principle, 101
product ($\prod$), 127
proof, 11, 12, 89
   by cases, 91, 257
   by contradiction, 93
   by contrapositive, 92
   by mathematical induction, 95–99
   direct, 92
proposition, 70
pseudocode, 2
pseudograph, 156

quantifier, 83
queue, 177
quotient, 3

range, 49

rational numbers, 30
real numbers, 30
recurrence relation, 115, 134
   boundary values, 115
   general solution, 125
   homogeneous, 122, 124
   initial conditions, 121
   linear, 125
   order of, 121
   second order, 122
   solution, 123
   solving a, 122
recursion
   base case, 109
   recursive step, 109
recursive, 60, 122
recursive algorithm for subsets, 208
recursive approach, 112, 134
recursive arithmetic, 187
recursive definition, 110
reflexive property, 56
relative error, 35
remainder, 3
remove element from a list, 128
repeat – until, 274
replace, 129
reverse, 119
root vertex, 176
rounding, 32, 35
rule of addition, 191
rule of complement, 191
rule of product, 190
rule of replacement, 81
rule of transitivity, 82
rules for
   elimination of connectives, 251
   introduction of connectives, 250
   quantifiers, 254
rules of logical inference, 81–82, 99, 250–255
Russell, Bertrand, 47

schema, 100
search, 142
search tree, 176–180
    breadth first, 178
    depth first, 179
semigroup, 230
sequence, 60–61
sequential composition, 264
set, 60
    notation, 42
set difference, 45
sexagesimal, 38
Sigma (Σ) notation, 111–112
sign, magnitude, 26
simplex algorithm, 141
singleton, 44
sorting, 144
source, 48
space complexity, 138
spanning tree, 175
specification, 6, 260
square matrix, 160
square root algorithm, 35
stamp problem, 95
state, 260
statement, 70
string, 18, 60–62
subgraph, 159
subset, 43
subtraction
    of integers, 23
    of rationals, 30
successor, 65, 188
sufficient, 77
sum, 110
sum of geometric series, 97
summation notation, 111
symbolic logic, 69
symmetric property of relation, 56
symmetric difference, 45

tail, 119
tally, 188
target, 49
tautology, 74, 79
termination, 145
ternary, 185
The Elements, 11
time complexity, 138
total function, 48
Towers of Hanoi, 133
trail, 163
transitive property of relation, 56
travelling salesman, 135, 216
tree, 166, 170–172
truncation, 35
truth tables, 71–77
two's complement, 26–30
type, 271

unary operation, 222
union of sets, 44
universal calculus, 69
universal set, 41

variables, 271
Venn diagram, 45
verification, 89
vertex (node), 155
vertex label, 159

walk of length n, 163
weakest precondition, 262
well-ordered, 59, 101, 221
well-ordering principle, 103
while loop, 265
window, 63
word, 60
word argument, 195
worst-case, 140